CHINA IN THE 21ST CENTURY

CHINA AND THE U.S.

COMPARISONS OF GREEN ENERGY PROGRAMS AND IRON AND STEEL PRODUCTION

CHINA IN THE 21ST CENTURY

Additional books in this series can be found on Nova's website under the Series tab.

Additional E-books in this series can be found on Nova's website under the E-book tab.

ENERGY POLICIES, POLITICS AND PRICES

Additional books in this series can be found on Nova's website under the Series tab.

Additional E-books in this series can be found on Nova's website under the E-book tab.

CHINA IN THE 21ST CENTURY

CHINA AND THE U.S.

COMPARISONS OF GREEN ENERGY PROGRAMS AND IRON AND STEEL PRODUCTION

JULIAN A. PAQUETTE
EDITOR

nova publishers
New York

For permission to use material from this book please contact us:
Telephone 631-231-7269; Fax 631-231-8175
Web Site: http://www.novapublishers.com

Library of Congress Cataloging-in-Publication Data

ISBN: 978-1-62948-389-4

Published by Nova Science Publishers, Inc. † New York

CONTENTS

PREFACE

China is the world's most populous country with approximately 1.3 billion people. It has experienced tremendous economic growth over the last three decades with an average annual increase in gross domestic product (GDP) of 9.8% during that period. This rapid economic growth has led to an increasing demand for energy, spurring China to add an average of 53 gigawatts (gw) of electric capacity each year over the last ten years to its power generation capabilities. China is also the world's largest producer and consumer of coal, with about half of its coal use being for electricity generation. In fact, coal provides over 70% of China's current electricity needs, and fuels much of the new power generation capacity being built. While many of these new coal plants are among the most technically advanced in the world, the burning of coal results in sulfur dioxide, nitrogen oxides, and particulate emissions which contribute to air pollution, and greenhouse gas emissions linked to global climate change. The current and potential future environmental consequences of burning coal are a major reason China has been actively seeking to increase its renewable energy capabilities. When current rates of use are considered, limited domestic reserves of coal, natural gas, and oil provide another impetus for change. However, China's announced intent to rely on domestic, sustainable solutions for its growing energy needs has led to a focus on developing "green" or renewable energy resources. In contrast to China, some argue that the United States does not have a comprehensive national policy in place for promotion of renewable energy technologies, with some observers saying that the higher costs of renewable electricity are not conducive to market adoption. However, for both countries, the reasons for increasing the use of renewable energy are diverse, and include energy security, energy independence, cleaner air, and more recently anthropogenic climate change, sustainability concepts, and economic development. Creating new, higher quality jobs could reasonably be said to be primary drivers of policy goals in both the United States and China.

Chapter 1 – China is the world's most populous country with over 1.3 billion people. It has experienced tremendous economic growth over the last three decades with an annual average increase in gross domestic product of 9.8% during that period. This has led to an increasing demand for energy, spurring China to add an average of 53 gigawatts (gw) of electric capacity each year over the last ten years to its power generation capabilities. China has set ambitious targets for developing its renewable energy resources with a major push of laws, policies, and incentives in the last few years. The wind power sector is illustrative of China's accomplishments, as installed wind power capacity has gone from 0.567 gw in 2003 to 12.2 gw in 2008, and China surpassed the United States in 2010 with over 41 gw of installed wind power capacity. Notably, however, approximately one-third of that capacity is not yet connected to the power grid. Plans already exist to grow China's wind power capacity to 100 gw by 2020. A similar goal exists for the solar photovoltaic power sector which China intends to increase generating capacity from 0.14 gw as of 2009 to over 1.8 gw by 2020. A hold on large and medium-scale hydropower project development has been lifted, with a virtual doubling of hydropower capacity planned. Most recently, China pledged ahead of the Copenhagen talks in 2009 that 15% of total energy consumption will come from non-fossil fuel sources by the year 2020. The 12th Five Year Plan will encompass 2011 to 2015, and will further formalize the link to green energy with specific deployment goals and investment. China recognizes that developing its domestic renewable energy industry and building its manufacturing capacity will help it meet energy demands at home and win advantages in future export markets. The key piece of legislation in recent years for advancing renewable electricity in China is the Renewable Energy Law of 2005. The law was designed to "promote the development and utilization of renewable energy, improve the energy structure, diversify energy supplies, safeguard energy security, protect the environment, and realize the sustainable development of the economy and society." Renewable energy is subsidized by a fee charged to all electricity users in China of about 0.029 cents per kilowatt-hour, and was originally based on the incremental cost difference between coal and renewable energy power generation. However, energy efficiency and conservation are officially China's top energy priority. These are considered the "low-hanging fruit" in the quest to reduce energy use and cut demand. Energy conservation investment projects have priority over energy development projects under the Energy Conservation Law of 1997, with government-financed projects being selected on "technological, economic and environmental comparisons and validations of the projects." China is the world's largest market for new construction, and new building standards have been in development since 2005 with national energy design criteria for residential buildings. In the power generation sector, many smaller, less efficient coal-fired power plants have been closed. In contrast to China, some argue that the United States does not have a comprehensive national policy in place for promotion of renewable energy technologies, with some observers saying that the higher costs of renewable electricity are not conducive to market adoption. However, for both countries, the reasons for increasing the use of renewable energy are diverse, and include energy security, energy independence, cleaner air, and more recently anthropogenic climate change, sustainability concepts, and economic development. Creating new, higher quality jobs could reasonably be said to be primary drivers of policy goals in both the United States and China.

Chapter 2 – Production of iron and steel is an energy-intensive manufacturing process. In 2006, the iron and steel industry accounted for 13.6% and 1.4% of primary energy consumption in China and the U.S., respectively (U.S. DOE/EIA, 2010a; Zhang et al., 2010). The energy efficiency of steel production has a direct impact on overall energy consumption and related carbon dioxide (CO_2) emissions. The goal of this study is to develop a methodology for making an accurate comparison of the energy intensity (energy use per unit of steel produced) of steel production. The methodology is applied to the steel industry in China and the U.S. The methodology addresses issues related to boundary definitions, conversion factors, and indicators in order to develop a common framework for comparing steel industry energy use. This study uses a bottom-up, physical-based method to compare the energy intensity of China and U.S. crude steel production in 2006. This year was chosen in order to maximize the availability of comparable steel-sector data. However, data published in China and the U.S. are not always consistent in terms of analytical scope, conversion factors, and information on adoption of energy-saving technologies. This study is primarily based on published annual data from the China Iron & Steel Association and National Bureau of Statistics in China and the Energy Information Agency in the U.S. This report found that the energy intensity of steel production is lower in the United States than China primarily due to structural differences in the steel industry in these two countries. In order to understand the differences in energy intensity of steel production in both countries, this report identified key determinants of sector energy use in both countries. Five determinants analyzed in this report include: share of electric arc furnaces in total steel production, sector penetration of energy-efficiency technologies, scale of production equipment, fuel shares in the iron and steel industry, and final steel product mix in both countries. The share of lower energy intensity electric arc furnace production in each country was a key determinant of total steel sector energy efficiency. Overall steel sector structure, in terms of average plant vintage and production capacity, is also an important variable though data were not available to quantify this in a scenario. The methodology developed in this report, along with the accompanying quantitative and qualitative analyses, provides a foundation for comparative international assessment of steel sector energy intensity.

In: China and the U.S.
Editor: Julian A. Paquette
ISBN: 978-1-62948-389-4

Chapter 1

CHINA AND THE UNITED STATES—A COMPARISON OF GREEN ENERGY PROGRAMS AND POLICIES*

Richard J. Campbell

SUMMARY

China is the world's most populous country with over 1.3 billion people. It has experienced tremendous economic growth over the last three decades with an annual average increase in gross domestic product of 9.8% during that period. This has led to an increasing demand for energy, spurring China to add an average of 53 gigawatts (gw) of electric capacity each year over the last ten years to its power generation capabilities.

China has set ambitious targets for developing its renewable energy resources with a major push of laws, policies, and incentives in the last few years. The wind power sector is illustrative of China's accomplishments, as installed wind power capacity has gone from 0.567 gw in 2003 to 12.2 gw in 2008, and China surpassed the United States in 2010 with over 41 gw of installed wind power capacity. Notably, however, approximately one-third of that capacity is not yet connected to the power grid. Plans already exist to grow China's wind power

---

* This is an edited, reformatted and augmented version of a Congressional Research Service publication, CRS Report for Congress R41748, prepared for Members and Committees of Congress, from www.crs.gov, dated March 30, 2011.

capacity to 100 gw by 2020. A similar goal exists for the solar photovoltaic power sector which China intends to increase generating capacity from 0.14 gw as of 2009 to over 1.8 gw by 2020. A hold on large and medium-scale hydropower project development has been lifted, with a virtual doubling of hydropower capacity planned. Most recently, China pledged ahead of the Copenhagen talks in 2009 that 15% of total energy consumption will come from non-fossil fuel sources by the year 2020. The 12th Five Year Plan will encompass 2011 to 2015, and will further formalize the link to green energy with specific deployment goals and investment. China recognizes that developing its domestic renewable energy industry and building its manufacturing capacity will help it meet energy demands at home and win advantages in future export markets.

The key piece of legislation in recent years for advancing renewable electricity in China is the Renewable Energy Law of 2005. The law was designed to "promote the development and utilization of renewable energy, improve the energy structure, diversify energy supplies, safeguard energy security, protect the environment, and realize the sustainable development of the economy and society." Renewable energy is subsidized by a fee charged to all electricity users in China of about 0.029 cents per kilowatt-hour, and was originally based on the incremental cost difference between coal and renewable energy power generation.

However, energy efficiency and conservation are officially China's top energy priority. These are considered the "low-hanging fruit" in the quest to reduce energy use and cut demand. Energy conservation investment projects have priority over energy development projects under the Energy Conservation Law of 1997, with government-financed projects being selected on "technological, economic and environmental comparisons and validations of the projects." China is the world's largest market for new construction, and new building standards have been in development since 2005 with national energy design criteria for residential buildings. In the power generation sector, many smaller, less efficient coal-fired power plants have been closed.

In contrast to China, some argue that the United States does not have a comprehensive national policy in place for promotion of renewable energy technologies, with some observers saying that the higher costs of renewable electricity are not conducive to market adoption. However, for both countries, the reasons for increasing the use of renewable energy are diverse, and include energy security, energy independence, cleaner air, and more recently anthropogenic climate change, sustainability concepts, and economic development. Creating new, higher quality jobs could reasonably be said to be primary drivers of policy goals in both the United States and China.

INTRODUCTION

China is the world's most populous country with approximately 1.3 billion people. It has experienced tremendous economic growth over the last three decades with an average annual increase in gross domestic product (GDP) of 9.8% during that period.[1] This rapid economic growth has led to an increasing demand for energy, spurring China to add an average of 53 gigawatts (gw) of electric capacity each year over the last ten years to its power generation capabilities.[2]

China is also the world's largest producer and consumer of coal, with about half of its coal use being for electricity generation. In fact, coal provides over 70% of China's current electricity needs, and fuels much of the new power generation capacity being built.[3] While many of these new coal plants are among the most technically advanced in the world,[4] the burning of coal results in sulfur dioxide, nitrogen oxides, and particulate emissions which contribute to air pollution, and greenhouse gas[5] emissions linked to global climate change.[6] The current and potential future environmental consequences of burning coal are a major reason China has been actively seeking to increase its renewable energy[7] capabilities. When current rates of use are considered, limited domestic reserves of coal, natural gas, and oil provide another impetus for change. However, China's announced intent to rely on domestic, sustainable solutions for its growing energy needs has led to a focus on developing "green" or renewable energy resources.[8]

China now has ambitious targets for developing both its hydropower and non-hydropower renewable energy resources with a major push of laws, regulations and incentives in the last few years. Development of large and medium-sized hydropower projects had previously been at a standstill with environmental impact and population displacement issues presenting major obstacles. [9] The wind power sector is illustrative of China's accomplishments, as installed wind power capacity has gone from 567 megawatts (mw) in 2003 to 12,200 mw in 2008. Plans already exist to grow China's wind power capacity to 100 gw by 2020. A similar goal existed for the China's solar photovoltaic (PV) power sector to increase capacity from 140 mw as of 2009 to 1.8 gw by 2020,[10] but now has been revised upward to as much as 20 gw by 2020. China also reportedly plans to virtually double its hydropower capacity to 380 mw by 2020. Developing its domestic renewable energy resources also provides a growth opportunity for China's domestic renewable energy technology industries, and can serve to increase its domestic manufacturing capabilities.

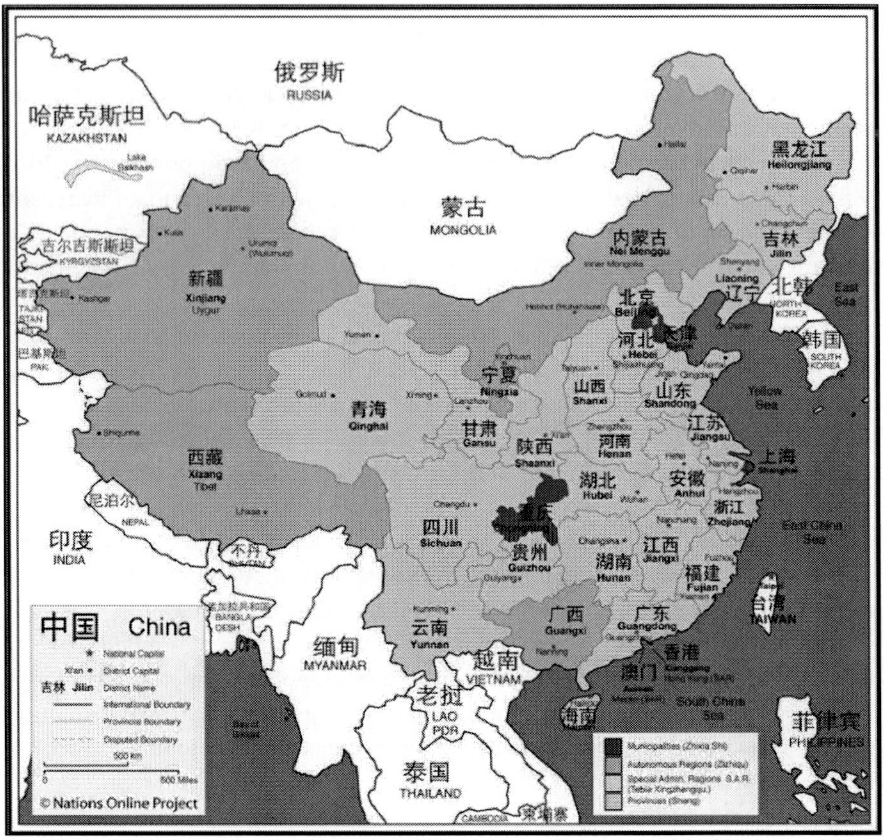

Source: http://www.nationsonline.org/maps/china-provinces-map-855.jpg.

Figure 1. Map of China. Administrative Regions and Provinces.

This report will look at the laws, programs, and policies encouraging development of wind, solar, and biomass power in the China and the United States as the major renewable electricity technologies common to both countries.

While hydropower is the most developed source of renewable electricity in both China and the United States, additional development of conventional hydropower is not currently a major focus of energy policy in the United States.

The terms "green" and "renewable" energy are used interchangeably for purposes of this report.

Existing Laws, Programs, and Incentives

Government incentives in the United States and China for renewable energy projects encompass a set of tools which generally depend on where a particular project is on a product development cycle. For example, direct subsides are more often applied earlier in the timeline of research and development (R&D) projects, while tax incentives are generally made available later in a cycle to aid manufacturing ventures or to encourage consumers to adopt the technologies and help to build demand.

China[11]

Most of the large industries in China (such as steel and petrochemicals) are comprised of state-owned enterprises which are essentially run by the national government. The five large power generation companies in China are also government entities: Huaneng, Guodian, Datang, Huadian, Zhongdiantou. Two utility grid companies exist—the State Grid Corporation covering the 22 provinces in northern and western China, and the Southern Power Grid in the five provinces in southern China (see *Figure 1).* For many years, the National Development and Reform Commission (NDRC) was the regulatory body in China with control over energy prices and project approval, and control over China's renewable energy development. As a result of recent reforms, energy policy strategy in China is now guided at the highest levels by the National Energy Commission (NEC) which is chaired by China's Premier Wen Jiabao.[12]

Large scale investment in clean energy technologies is a relatively recent undertaking for China, with air pollution concerns prompting the first forays into clean energy development. More recent attention to mitigating the perceived effects of global climate change have provided additional momentum.[13] Policies for encouraging renewable energy in China are largely driven by the central government, and enacted through national and provincial and local government programs. China led the world in 2010 in renewable energy investment, spending $54.4 billion, with the United States third in clean energy spending, investing $34 billion.[14]

Clean Energy Research and Development Programs[15]

A series of programs promoting research and development of renewable energy technologies were established by China in the last quarter of the past century under the Ministry of Science and Technology. These included the following:

- *Key Technology R&D Program:* Initiated in 1982 to address major science and technology issues in economic and social development, this was China's first national R&D program supporting innovation for environmental pollution control and efficient resource utilization for energy and water. Almost $1 billion was invested between 2001 and 2005.[16]
- *863 Program:* Known more formally as the "National High-Tech Development Plan," the 863 Program was created in March 1986 to develop a wide range of technology fields. The program focuses on boosting innovation in strategic high technology sectors so that China can gain a foothold in world markets. Its initial objective was to make China independent of financial obligations for foreign technologies, and to diversify research efforts away from purely military themes to civilian and dual-use technologies such as satellites, computers, robotics, biotechnology, energy and space exploration. The program invested $3 billion in research from 2001 to 2005, and another $585 million was approved in 2008 jointly for the 863 and 973 R&D programs.[17]
- *973 Program*: The National Basic Research Program focuses on fundamental, basic research and thus complements the 863 Program. Energy and sustainable development have been key features of the 973 Program since its founding at the third meeting of the National Science and Technology Committee in 1997. The program funded 382 projects between 1998 and 2008, with a total investment of $1.3 billion.[18]
- *Five Year Plans:* The Five Year Plans are the guide to China's economic growth. The 10th Five Year Plan (for 2001 to 2005) budgeted $2.4 billion for the implementation of 12 "mega-projects" based on the 863 and National Key Technologies Programs aimed at achieving significant technological breakthroughs for China's industries. The 11th Five Year Plan (for 2006 to 2010) identifies energy technologies as a focus of the 863 Program, with hydrogen, fuel cells, energy efficiency, clean coal and renewable energy a focus of $172 million in funds. The 11th Five Year Plan has also made utility-scale renewable energy and new energy development the 973 Program's main focus.[19] Renewable energy is increasingly being linked to China's future wellbeing—both economically and environmentally.

The 12th Five Year Plan will encompass 2011 to 2015 and was formally announced in March, 2011. Energy efficiency and renewable energy continue as a focus of China's government as part of seven new "Strategic Emerging Industries."[20] China will likely follow the plan with specific investment goals (through preferential tax, fiscal and procurement tools) in biotechnology, new energy (nuclear, solar, wind , biomass), High-end Equipment Manufacturing, Energy Conservation and Environmental Protection (energy efficiency, advanced environmental protection, recycling), Clean-Energy Vehicles, New Materials, and Next-Generation Information Technology. These industries are expected to become the "backbone of China's economy" in the near future, offering inroads to global markets.[21] China has pledged in the 12th Five year Plan to cut energy consumption per unit of GDP by 16% and cut carbon emissions by 17% in the period from 2011 to 2015.[22]

Key Legislation Promoting Renewable Energy

Renewable energy had been encouraged by a number of early laws in China, but these were intended more for rural development in distributed generation schemes. Many laws governing renewable energy do not have a lot of details prescribing how the legislation should be implemented. The laws generally state goals or what has to be accomplished, and lay out a framework. The details on how goals will be achieved are determined later, usually by the NDRC.

The main law governing China's power industry is the *Electricity Law of 1995*.[23] It was the first law to place legislative controls on the utility industry, centralizing control of existing regulatory agencies and restructuring state-owned electricity companies. The law was designed to promote the safe development of the electricity industry, and identified renewable energy as a means to develop electrification especially in rural areas.[24]

The *Energy Conservation Law of 1997* was enacted by the Standing Committee of the National People's Congress to guide the use of energy resources, promote energy-saving technologies, and protect the environment. The General Provisions state that "Energy conservation is a long-term strategy for national economic development," and the "state shall encourage and support research and popularization in the science and technology of energy conservation." While energy development and conservation are both being pursued, energy conservation investment projects are given priority over energy development projects under Article 10, with projects under the central government being selected on "technological, economic and environmental comparisons and validations of the projects." Article 11 then directs the State

Council and provincial governments to "allocate energy conservation funds from both capital construction and technological restructuring investment funds to support rational energy utilization and development of new and renewable sources of energy." Article 38 of the Law requires government "at all levels" to use renewable energy resources sustainably, especially for rural areas. Article 39 focuses on improving energy efficiency in a variety of applications:

> The State encourages the development of the following universal energy conservation technologies: (1) promote the wide use of cogeneration of heat and power and district heating, increase the utilization rate of heat and power units, develop heat-cascading technology, combined heat, power and cooling technology and combined heat, power and coal gas technology, and increase the efficiency of thermal energy application in an all-round way; (2) gradually achieve more-efficient operation of electric motors, fans, pumping equipment and systems; develop adjustable speed motor drives for energy conservation and electric-electronic power saving technology; develop, produce and disseminate the use of high-quality and low-cost energy-efficient appliances and equipment; and increase the efficiency of electric power; (3) develop and disseminate the use of clean coal technologies, including fluidized bed combustion, smokeless combustion, and gasification and liquefaction, that are suited to domestic coals in order to increase coal utilization efficiency; and (4) develop and disseminate other universal energy-efficient technologies that are proved mature and yield remarkable benefits.

Article 40 directs all trades and professions to seek and disseminate energy-efficient technologies or solutions, and to replace outdated technologies and equipment. [25]

The Energy Conservation Law is generally written to consider sustainability while feeding the growth of the economy. Conservation and energy efficiency measures are seen as ways to lower the financial and environmental costs of funding China's growing economy. Demand-side management is also a key feature of China's 2007 Climate Change Program[26] in reducing the energy intensity[27] of the economy through:

> the development of specific energy conservation plans, the adoption and implementation of technology, economic, fiscal and management policies in favor of energy conservation, the development and application

of energy efficiency standards and labeling, the encouragement of R&D, demonstration and diffusion of energy-saving technologies, the importing and absorbing of advanced energy-saving technologies, the creation and employment of new energy conservation mechanisms, and the promotion of key energy conservation projects.[28]

China's 11th Five Year Plan requires state-owned enterprises to reduce energy intensity by 20% overall from 2005 levels. Meeting this goal may prove to be a challenge. While China has been able to reduce energy intensity 14.4% from 2005 to 2009, there was an increase of 3.2% in energy intensity during the first quarter of 2010.[29]

China's *Law for Prevention and Control of Air Pollution of 2000* also looks to renewable energy as a means of preventing atmospheric air pollution. The law encourages the support and development of clean energy technologies for "solar energy, wind energy and water energy."[30]

However, the key piece of legislation in recent years for moving renewable energy deployment forward is the *Renewable Energy Law of 2005*.[31] The law was based on goals to "promote the development and utilization of renewable energy, improve the energy structure, diversify energy supplies, safeguard energy security, protect the environment, and realize the sustainable development of the economy and society." The law has several key elements:

- Allows for middle- and long-term national targets to be set for the total volume of renewable energy development (Article 7);
- Mandates connection with the grid and the purchase of electricity from licensed renewable energy generators (Article 14);
- Makes available preferential loans with subsidized interest rates (Article 25), and granted tax benefits (Article 26) for renewable energy projects.

While under the provisions of the law, the "energy authorities of the State Council" are to implement its provisions both locally and nationally, the NDRC developed a *Medium and Long-Term Development Plan for Renewable Energy* (MLTPRE) in 2007 to implement the law. The MLTPRE established national deployment goals by technology to reach renewable energy. The "Guiding Principles" of the MLTPRE state that it focused on hydropower, wind, solar, and biomass energy development and deployment, coordinating renewable energy development with economic, social, and environmental objectives. Overall, the MLTPRE aimed at raising the share of renewable energy to 10% of total primary energy consumption by the end of 2010, and 15% by 2020. The cost of renewable energy development and deployment in excess of conventional power (i.e., coal) is to be socialized by passing the expense to all customers as a surcharge to the retail price of power.

Mandates for market share under the MLTPRE require areas of China covered by large scale power grids to have non-hydropower renewable energy account for 1% of total power generation by 2010, and at least 3% by 2020. Power generators with installed capacities over 5 gw will be required to have non-hydropower renewable energy of 3% of total capacity by the end of 2010, and at least 8% by 2020.[32]

The national government recently added an incentive for the grid companies to connect to renewable energy projects. The Renewable Energy Law was amended in 2009 to require electricity grid companies to buy all the electricity produced by renewable energy generators. Power companies refusing to comply will be fined an amount up to twice the "economic loss" of the renewable energy producer.[33]

China's pledge for 15% of total energy consumption to come from non-fossil fuel sources by the year 2020, made ahead of the 2009 climate talks in Copenhagen,[34] was reiterated by the National Energy Administration[35] at a Beijing national work conference in January 2011.[36] Increased targets for specific renewable energy technology deployment will likely be a significant part of goals in the 12th Five Year Plan.

Renewable Energy Focus

China's renewable energy development goals are shown in *Table 1.* The major renewable energy technologies are described in the following paragraphs. Since China considers hydropower to be renewable energy, it is also included in the section.

Biomass

Biomass as a source of grid-connected electric power in China has yet to realize its full potential, with inefficiencies in transporting biomass feedstocks to centralized locations being a major hurdle. Biomass electric production was estimated to have reached 5,500 mw in 2010, and is expected to increase to 30,000 mw by 2020.[37] Biomass is widely used in China's rural areas where over 80% of the population live.[38]

Table 1. China—Renewable Energy Deployment Targets (Capacities by Year)

Renewable Energy Source	2005	2010	2020
Hydropower	117 gw	200 gwa	430 gwd
Biomass Power	2 gw	5.5 gw	30 gwd
Wind Energy	1.26 gw	42 gwa	150 gwd
Solar PV	0.07 gw	0.6 gwa	20 gwd
Bioethanolb	1.02 million tons	2 million tons	10 million tons
Biodieselc	50,000 tons	200,000 tons	2 million tons

Source: Compiled by CRS from various reports, including the Medium and Long-Term Development Plan (2007) from the National Development and Reform Commission (People's Republic of China). See http://en.chinagate.cn/reports/2007-09/13/content_8872839.htm.

[a] Actual installed capacity for 2010.

[b] Non-food grains used as feedstock.

[c] From waste and edible oil sources.

[d] Reported new target.

When biomass is used in biogas digesters, the resulting biogas mixture can be up to 70% methane.[39] Biogas produced from livestock manure is a major resource in rural areas for household cooking and heating.[40] Biogas methane can also be used to fuel engine generators and to produce electricity for households. Larger biogas projects from collectives can operate cogeneration facilities providing thermal energy for heating or hot water.

Biomass-fueled electricity is generally considered to be carbon-neutral, but biomass is a very small part of China's overall centralized electricity production. Development of biomass projects connected to the grid are expected to be limited to areas with abundant biomass resources in order to promote direct-fired biomass electric power generation plants. As of 2006, biomass electric power capacity was about 2,200 mw.[41] Biomass power projects are eligible for a feed-in tariff[42] (FIT) which was raised to $0.051 per kilowatt-hour in 2008.[43] A production subsidy was authorized of $19 to $22 per ton of biomass pellets produced from agricultural or forestry residues.[44] Biomass energy is also viewed as a part of the solution to arrest desertification in China, with programs to plant willow trees springing up in affected regions. Willow trees grow quickly, and are harvested for energy. For example, a power plant in Inner Mongolia burns as much as 200,000 tons of willow annually, producing 210 million kwh of electricity, thus displacing power from coal-based generation. The ashes left over from the combustion process can be used for fertilizer.[45]

With China's domestic oil reserves dwindling, biofuels are considered a supplement to enhance China's transportation energy supplies, and reduce air emissions. Over $290 million was allocated for research, development, and demonstration of biofuels. China is the world's third largest producer of ethanol, following the United States and Brazil. While biofuels can be made from a variety of biomass sources, China's MLTPRE focuses on using marginal lands and non-food crops for biofuels production,[46] with specific targets for bioethanol and biodiesel (see *Table 1*). It should also be noted that China has become a net food importing country which makes food security a priority, and is a likely reason for China's focus on non-food crop sources of bioethanol.[47]

Hydropower

China is now the world's largest hydropower producer with over 200 gw of installed capacity.[48] That capacity is expected to increase to 290 gw by 2015, according to goals announced by China's National Energy Administration, with major new dams to be installed on the Huanghe, Jinshangjiang,Yalongjhiang, Daduhe, Nujiang, and Lancangjiang rivers.[49] China's overall potential for 'technically developable" hydropower is estimated to be 540 gw, according to the MLTPRE. Over 400 gw of that capacity seen as "economically developable" with 70% of the best resources for hydropower development in southwest China.[50] With the lifting of the effective "ban" on large and medium-scale hydropower development,[51] China reportedly now plans to develop approximately 430 gw of hydropower capacity by 2020.[52] Past dam-building booms created many water resource and environmental problems, but China's government has passed a variety of water-protection laws and regulations over the last 20 years which it hopes will reduce environmental impacts of the current dam-building cycle.[53]

Solar

China is the world's largest producer of solar photovoltaic (PV) panels, accounting for 72% of new PV manufacturing capacity globally.[54] Installed capacity in China has grown more slowly than production, with a total of 0.6 gw of solar PV installed as of 2010.[55] However, this represents a huge growth since 2008 when China had installed about 140 mw of PV capacity.[56] With the global economic downtown came a slowdown in orders for PV panels. China began to look at developing the solar power domestic market.

China's initial goal for solar power was established in the 2007 MLTPRE at 1.8 gw, but this target is in the process of being revised upward to perhaps 20 gw.[57] A "Golden Roofs" initiative announced in March 2009 provides a subsidy of $2.93 per watt for roof-mounted PV systems over 50 kilowatts (kw) which could cover over half of a system's installation cost. A feed-in tariff of $0.16 per kilowatt-hour (kwh) was also established for PV power projects at the same time. Encouragement for larger utility scale solar projects was announced in July 2009 under the "Golden Sun" program, which provides for up to 50% of project costs (including transmission or distribution lines to connect to the grid), and up to 70% of such costs for projects in more remote areas (such as the Western Region). The Golden Sun program is for projects of 300 mw capacity and above, which are in service for a minimum of 20 years.[58]

Provinces also provide local incentives for solar development. For example, the Jiangsu provincial government established a FIT for solar power from ground-based solar farms, rooftop, or building integrated PV systems installed in 2009 with respective rates of $0.31, $0.54, and $0.63 per kwh of electricity generated.[59] There has since been discussion of establishing a national feed-in tariff for solar power projects. However, solar power may be in competition for national government funding with wind power (which is seen by the central government leadership as a better option for China) for such consideration, making a solar FIT unlikely in 2010.[60]

Pilot plants for large scale concentrating solar power facilities have been proposed for Gansu (300 mw) and Shaanxi (92 mw), with discussions for a demonstration plant in Inner Mongolia (50 mw).[61]

Wind

China installed 16 gw of wind power capacity in 2010, lifting its overall total to almost 42 gw, making it the world's leader of installed wind power capacity. Even so, China lags behind the United States in total energy produced from wind power with 50 billion kwhs generated. China's grid-connected capacity lags behind installed capacity by over 30%.[62]

Domestic wind power turbine technology and electricity production have grown tremendously in China since the turn of the century. National government support for wind power in China began in 2001 with a 50% cut in value-added taxes for power generated by wind. This was followed in 2003 with a push for wind power development from the Chinese government with the introduction of a tender process for award of concessions for wind power projects.[63]

The process for wind farm development usually begins with the NDRC conducting wind resource assessments for prospective areas prior to arranging for a concession for a wind power project. Projects below 50 mw do not require competitive bidding, and so may be developed by local authorities. Wind power projects over 50 mw are approved by the NDRC, which also sets prices for the electricity generated by these projects. While wind projects under 50 mw can be approved by local governments, prices for wind power are generally subject to final approval by the NDRC. [64]

The regional grid power company would enter into a long-term power purchase agreement to buy electricity from the selected bidder over the life of the wind project, with the national government guaranteeing the power purchase. The bidding process would also determine the in-grid tariff, with the agreement specifying how much electricity the bidder would provide to the grid. The goal of the program was to achieve economies of scale by producing a large capacity and thereby produce a low price for grid-connected wind power. Additionally, it was thought that foreign companies would be attracted by the long-term purchase power agreement to invest in China's wind energy sector.[65]

Mixed results came from the tenders received, with many being structured on impractically low power prices. Winning bids were thus often too low to make the projects economically viable. This prompted the government to change the weight of power prices in its process of evaluating the bids.[66]

The 2005 Renewable Energy Law established a purchase system for renewable electricity, but the process of requesting bids for tenders continued for grid-connected wind power projects. In 2007, a target of 15% of China's total energy consumption from renewable energy was set for 2020. The law required the grid company to purchase the full amount of power generated by wind power projects with the tariff for all projects being set by the winning bid.[67]

In 2009, the NDRC replaced the public bidding process and instituted FITs for wind power, scaled according to the available wind resource and construction conditions in the various regions of China.[68] As more projects were installed, the understanding of the dynamics between localized wind resources and resulting power production led to more rational prices for wind power.

As of 2009, China's Meteorological Administration estimated China's developable wind power resources at over 250 million kilowatt-hours.[69] By the end of 2009, China's installed wind power capacity had reached 25.8 gw.[70] Officials in China expect to surpass the official MLTPRE target of 30 gw for installed wind power by 2020, with estimates of 150 gw of wind power capacity predicted. The revised expectations have yet to be formally endorsed by the NDRC.[71] However, China's wind resources and increasing demand for electric power are now being matched by the industrial capability to build wind turbines and solar panels which can meet the resource potential.

Offshore wind is another mostly untapped resource in China. China's offshore wind power potential is estimated at more than 750 million kilowatt-hours, which is more than twice the estimate of exploitable wind power resources onshore. As with onshore wind, offshore concessions would be put up for tender offers with price offers to send electricity to the grid. Developers must be Chinese-funded enterprises, or Sino-foreign joint ventures with majority Chinese ownership. The process of establishing concession areas has already begun with China's National Energy Bureau, and the State Oceanic Administration jointly issuing an "Interim Measure" in 2010 concerning regulations for developing offshore capacity.[72]

Incentives, Subsidies, and Procurement Programs

Financial support for renewable energy in China involves subsidies, tax polices, pricing mechanisms, and a reward scheme for green production. Subsidy support is extended to overhead costs of programs (i.e., administrative, operational, and other expenses for government renewable energy agencies), renewable energy technology research and development, and provincial or local electrification projects. Tax incentives can come from the central or local governments, and can be technology specific. Pricing for renewable energy is not standardized, and is set by contracts negotiated between projects and utilities.[73]

Renewable energy is subsidized by a rate fee charged to all electricity users in China of about 0.029 cents per kwh.[74] Electricity customers in China pay rates according to customer class.[75] The fee was originally based on the incremental cost difference between coal and renewable energy (estimated in China at $0.044 to $0.059 per kwh).[76] The fee goes to the companies which operate the electricity grid and must buy the renewable power from project developers. The fee for industrial users of electricity recently doubled in 2009 to about 0.8% of their electricity bill.[77]

As China's technological capabilities improve, it will use more of its own industrial capacity to meet domestic clean energy needs, and rely less on imported equipment. China has embraced an array of incentives, subsidies, and procurement policies to encourage such development. Interest rates as low as 2% for bank loans enable the financing of renewable energy projects.[78]

Preferences were established with the Government Procurement Law of 2002, which specified the government shall procure domestic goods, construction, and services, unless availability or other conditions exist to impair such procurement.[79]

Application of domestic content rules for renewable energy projects were formalized in 2005 by the NDRC's "Notice of Requirements for the Administration of Wind Power Construction." At least 70% of Wind Power equipment is required to be produced in China. The determination of domestic content is based on the percentage of total components in a wind turbine manufactured and assembled in China, regardless of the level of Chinese ownership in the producing factory. Wind farms which fail to comply do not receive approval for construction permits. Authorities in China are reportedly considering dropping the domestic content requirement, when objectives of establishing the domestic wind turbine industry and supply chain are considered met.[80] Under the Wind Power Concession Project, the NDRC is "overseeing construction" of a series of wind farms of at least 10 gw generating capacity.[81]

Financial incentives for other renewable energy projects in China are available from both the national and provincial governments. The central government offers an investment subsidy of 50% for solar power projects under the Golden Sun program. For biomass power projects, a $0.04 per kwh subsidy is offered, along with incentives such as risk reserves and tax breaks, and the government has established at least one joint venture to demonstrate and deploy biomass power technology at a reported 40 plants.[82]

Incentives for non-food sourced biofuels production are available to farmers and biofuel producers. Ethanol production in 2006 was 1.56 million tonnes compared to 0.19 million tones for biodiesel, with subsidies for ethanol at $115 million and no subsidies for biodiesel in that year. In 2007, flexible subsidies were made available to biofuels producers to make up for losses on crops due to low crude oil prices (though the Chinese government wants to encourage enterprises to reserve funds to offset such risks). Farmers were authorized a subsidy of $405 for each hectare of forest used for biofuels production, and a subsidy of $365 for each hectare growing biofuel crops.[83] China controls transportation fuel prices, and sets the price of ethanol at approximately 91% of the price of gasoline. The five licensed producers of ethanol in China are eligible for support in the form of direct output-linked subsidies, tax exemptions, and low-interest loans, and they are the beneficiary of mandatory blending programs for ethanol with gasoline in ten provinces. Contrary to the highly controlled ethanol industry, the biodiesel industry is dominated by small scale producers who use waste cooking oil and animal fats as feedstock. Producers sell biodiesel direct to users without taxation or direct fuel subsidies. Total support for ethanol and biodiesel is expected to rise to $1.2 billion by 2020, excluding the subsidy support to farmers mentioned earlier.[84]

Under the economic stimulus plan designed to help China recover from the current global financial crisis, the national government allocated over 210 billion yuan (about $31 billion), or 5.3% of its entire stimulus package, for environmental protection and energy conservation.[85]

China provides substantial domestic subsidies to its green energy industries in support of its ambitious deployment goals. However, in the wake of complaints in the United States by the United Steelworker's Union and others that China was illegally subsidizing its wind power and solar PV exports in violation of World Trade Organization (WTO) obligations, the U.S. Trade Representative announced an investigation in 2010 of China's support for makers of wind power, solar energy, advanced batteries, and energy-efficient vehicles.[86] China's export restrictions on rare-earth elements[87] (REEs) were described by the United Steelworkers Union as an example of China's unfair trade practices. China produces 97% of the world's REEs and charges a 10% export tariff. China also cut its exports of REEs by 40% in July 2010, causing global demand to exceed supplies for the first time.[88] The WTO reportedly issued an interim report in February 2011, which stated that Chinese restrictions on its exports of nine raw materials were inconsistent with WTO rules, which may induce WTO members to bring a similar case against China over its export restrictions on REEs.[89] China's government has denied the allegations of unfair trade practices, pointing to proposed federal incentives for the U.S. clean energy industry as comparable subsidies.[90]

United States

Some argue that the United States does not have a comprehensive national policy in place for promotion of renewable energy technologies. Federal policies exist providing corporate tax incentives for renewable electricity, but these are generally authorized for short periods and must be periodically reauthorized. Even when the programs are authorized for longer terms, they are not always fully funded in appropriations legislation. Most federal grant and loan programs are short-term in funding duration, with the programs authorized by the American Recovery and Reinvestment Act of 2009 being an example. An exception, however, is biofuels which do have significant federal government support in the form of the Renewable Fuel Standard (RFS). The RFS mandates minimum goals for blending quantities of renewable fuels with gasoline and diesel through 2022.[91]

Much of the U.S. renewable energy installed capacity is a result of state deployment initiatives[92] rather than federal programs, with 30 states having a renewable portfolio standard (RPS) in place to encourage deployment.[93] However, the availability of federal tax incentives have aided deployment in recent years, with the Investment Tax Credit[94] (ITC) being key to many corporate decisions to invest in renewable energy.

Clean Energy Research and Development Programs

The history of energy research and development in the United States in the closing quarter of the last century can be characterized as being driven by energy prices, causing shifts in the direction of policies, programs, and levels of program funding. During this period, the U.S. Department of Energy (DOE) has been the key agency involved in energy R&D in the United States according to a 2001 report from the National Research Council:

> From 1978 through 1999, the federal government expended $91.5 billion (2000 dollars) on energy R&D, mostly through DOE programs. This direct federal investment constituted about a third of the nation's total energy R&D expenditure, the balance having been spent by the private sector. Of course, government policies—from cost sharing to environmental regulation to tax incentives—influenced the priorities of a significant fraction of the private investment. On balance, the government has been the largest single source and stimulus of energy R&D funding for more than 20 years.[95]

DOE continues to be responsible for the largest share of federal energy-related research dollars, and administers the national laboratories and technology centers which are key to the execution of U.S. national energy research strategies.

The modern history of energy research can be traced to the 1977 Department of Energy Organization Act (P.L. 95-91) which dismantled the predecessor agency (the Energy Research and Development Administration) and created the agency.

Key Legislation Promoting Renewable Energy

The *National Energy Act of 1978* followed the formation of DOE, and was largely focused on conservation of fossil fuels in reaction to the 1973 energy crisis. The Public Utility Regulatory Policies Act (PURPA) (P.L. 95-617) encouraged generation of electricity from renewable sources by requiring electric utilities to purchase electricity from qualifying small power and cogeneration facilities.[96]

It was the *Energy Security Act of 1980* (ESA) (P.L. 96-294) which largely brought renewable energy and renewable technologies into the forefront of public policy. ESA consisted of six major acts[97] and provided funding for research in areas such as renewable energy and biofuels.

The *Energy Policy Act of 1992* (EPACT) (P.L. 102-486) was wide-ranging legislation addressing topics of energy efficiency and conservation, natural gas supplies, alternative fuels and alternative fuel vehicles. EPACT set goals for energy management,[98] and authorized subsidies for wind and other alternative energy technologies (i.e., the Renewable Energy Production Incentive discussed later in this report). EPACT also established the Production Tax Credit[99] (PTC).

The *Energy Policy Act of 2005* (EPACT5) (P.L. 109-58) continued the focus on energy supply and demand policies. EPACT5 extended the PTC for wind and qualifying biomass technologies. The law also authorized funds for developing renewable energy technologies and loan guarantees for renewable energy deployment, and required electric utilities to offer net metering on request to customers. EPACT5 also created the Renewable Fuel Standard with requirements for blending 7.5 billion gallons of renewable fuel with gasoline by 2012.

This was followed by the *Energy Independence and Security Act of 2007* (EISA) (P.L. 110- 140) which was largely focused on energy security and energy efficiency. EISA also provided funds to accelerate R&D for renewable energy particularly solar and geothermal power, and energy storage technologies. EISA also extended the RFS program with a mandate to blend 36 billion gallons of renewable fuel with gasoline and diesel fuel by 2022.

The *American Recovery and Reinvestment Act of 2009* (ARRA) (P.L. 111-5) was enacted as a stopgap measure in response to the current financial crisis to aid economic recovery.

> More than $45 billion was appropriated for energy efficiency and renewable energy programs across federal government programs, most of which was to be obligated before the end of FY2010. Almost $8 billion

> was provided for energy and other R&D programs, $2.4 billion for energy technology and facility development grants, and $14 billion for electric power transmission grid infrastructure development and energy storage development (including $6 billion for loan guarantees). Another $14.1 billion was provided for renewable energy tax incentives, with an additional $2.3 billion for energy efficiency tax incentives.[100]

ARRA also included a temporary provision allowing projects eligible for the Investment Tax Credit to receive a grant covering 30% of project costs.

Renewable Energy Focus

The principle U.S. renewable energy sectors are described in the following paragraphs. Hydropower is not viewed by all U.S. experts as renewable energy, but since China has significant goals for hydroelectric development, a summary of U.S. hydropower is included.

Biomass[101]

Biomass for electric power is arguably the most conventional of all renewable electricity technologies. Approximately 23.5% of non-hydropower renewable electricity produced in the United States came from biomass in 2010.[102] Biomass combustion is a relatively mature technology but it is not widely used and is generally not very efficient unless it is used in a combined heat and power application. Large scale co-firing of biomass with coal is a higher efficiency, lower per unit cost application, and is categorized usually under coal power generation. Technologies for biomass gasification could result in higher efficiencies when used to produce synthesis gas or hydrogen for heat and/or power production. Demonstration and deployment of newer industrial gasification technologies is needed to scale-up plants and provide economical designs with high degrees of availability. Wood-burning stoves and solar water heaters are the most common residential renewable energy applications. With wood and biomass electric power net summer generating capacity reported at 7.6 gw for 2010, DOE estimates that 17 gw of biomass capacity could be available by 2030.[103] The RFS is a significant factor in biofuels policy which has largely focused on corn-based ethanol[104] but is shifting to cellulosic and other advanced feedstocks.[105] Increasing scrutiny on the effects of ethanol production on global food prices and production practices has pushed researchers to focus more on non-food cellulosic sources production of ethanol.[106]

Hydropower

While only about 2,400 of the existing 80,000 dams in the United States are used for hydropower, many of the non-power dams have a significant unharnessed hydropower potential. A DOE study in 2003 indicated that enhancing existing hydro facilities, by adding turbines to dams without any hydro capacity or enhancing existing structures (increasing capacity or adding turbines), is relatively inexpensive and could present a further opportunity for hydropower development.[107]

Most of the better sites for larger, conventional hydropower projects in the United States have already been developed. However, DOE's Idaho National Laboratory assessed the potential for developing small and low-head hydroelectric generation in the United States.[108] A set of feasibility criteria was established for "developable resources, and identified approximately 5,400 with the potential for small hydro projects (i.e., providing between 1 MW and 30 MW of annual mean power). DOE estimated these projects (if developed) could result in a greater than 50% increase in total U.S. hydroelectric power generation.

Solar

As of 2010, the net summer grid-connected solar power industry in the United States was approximately 2.9 gw representing both solar PV and solar thermal capacity. DOE estimates that approximately 16 gw could be available by 2020.[109] Solar power, like wind power, is considered a variable resource but solar power technologies can generate most power at peak energy demand times when the weather is hot and sunny. Concentrating solar power thermal plants with heat storage capacity are being increasingly considered for large central station generating plants in the sun-rich areas of the western United States.

Solar thermal hot water heating is a small but growing application in the United States whose deployment prospects may be enhanced by a recent focus on energy efficiency and conservation. For the solar power industry, the key federal incentive of recent years has been the Investment Tax Credit which allowed businesses to invest in solar power projects and receive a tax credit for up to 30% of the expense. As a short-term remedy to the almost annual reauthorization quest for the ITC, the Emergency Economic Stabilization Act of 2008 extended the 30% solar investment tax credit for eight years to 2016, and removed the prohibition against utility company use of the ITC, thus allowing them to take advantage of the credit.

Wind

The net summer wind power capacity in the United States was approximately 38 gw in 2010.[110] However, DOE estimates that domestic wind power could reach a capacity of 61 gw or greater by 2030.[111] A major federal government incentive for wind power has been the Production Tax Credit (PTC), which originated in the Energy Policy Act of 1992 as aid to facilities in operation. Currently, an income tax credit of $0.021 per kwh is available for electricity produced from utility-scale wind turbines under the PTC. However, the PTC has expired three times in the last decade only to be restored. The importance of the PTC to the industry is apparent as installations of wind power have consistently fallen in the year following the lapse of the tax credit.[112] The American Recovery and Reinvestment Act of 2009 (ARRA) (P.L. 111-5) extended the Section 45 Production Tax Credit "placed in service" date for wind to the end of 2012, and allowed PTCeligible facilities placed in service from 2009 and 2012 to choose a 30% ITC in place of the PTC, or to receive a 30% grant.

Offshore wind power in the United States is a fledgling industry, having just received federal authority in 2010 to go ahead with the first U.S. offshore wind farm in Nantucket Sound, off the Massachusetts coast. Known as the Cape Wind project, it will involve 130 turbines (from the German firm, Siemens AG) with a total capacity of 420 mw.[113] The overall potential for U.S. offshore wind power production capacity was estimated at 908 gw in 2005.[114]

Incentives, Subsidies and Procurement Programs

Government incentives, subsidies, and procurement requirements exist at both the federal and state government levels in the United States. The federal government is the largest single consumer of energy in the United States, and has defined procurement goals for the use of renewable energy:

> EPACT5 required federal agencies to increase their purchase of renewable energy to a minimum of 7.5% of overall energy purchases by 2013. Agencies receive double-credit for renewable energy generated on their facility sites. Presidential Executive Order (EO) 13423 required that at least one-half of the EPACT renewable energy requirement come from "new" (i.e., put in service after January 1, 1999) renewable energy sources, preferably sited on agency property for agency use. EO 13423 also allowed agencies to use new "nonelectric" renewable energy sources to meet the requirement for new renewable energy. Examples of non-electric renewable energy include thermal energy from solar ventilation

pre-heat systems, solar heating and cooling systems, solar water heating, ground source heat pumps, biomass-fueled heating and cooling, thermal uses of geothermal and ocean resources. However, these non-electric renewable energy sources cannot apply to meeting the EPACT renewable federal electricity purchase requirement. In 2010, an agency could use nonelectric renewables equal to 2.5% of its electricity to satisfy EO 13423, and then use old renewable energy sources for 5% of its use to satisfy EPACT, for a total equivalent of 7.5% of its electricity use from renewable energy.[115]

The Renewable Energy Production Incentive (REPI) was established by EPACT to provide incentive payments for new projects generating and selling renewable electricity. Eligible renewable energy technologies include solar, wind, and biomass (excluding municipal solid waste). The payment of $0.015 per kwh (in 1993 dollars, indexed for inflation) was for the first 10 years of a facility's operation, but is subject to availability of annual appropriations in each federal fiscal year of operation.[116] REPI was reauthorized by EPACT5 for FY2006 through FY2026. REPI program funding is determined under the DOE budget process, and employs a tiered decision process as to which projects have priority for payments.[117]

Under the Food, Conservation and Energy Act of 2008 (2008 Farm Bill) (P.L. 110-234), the U.S. Department of Agriculture runs a grant program for developing renewable energy and energy efficiency projects under the Rural Energy for America Program (REAP). REAP is intended to encourage the adoption of renewable energy and energy efficiency technologies by rural small businesses and farmers largely through the competitive issuance of grants and loan guarantees. The program also funds energy audits and provides other renewable energy technology assistance. The 2008 Farm Bill provided for base funding of REAP at $55 million for FY2009, $60 million for FY2010, $70 million for FY2011, and $70 million for FY2012.[118]

DISCUSSION

With the process of urbanization continuing in China, a further 200 million to 300 million people are expected to move to urban areas over the next 20 years. This shift in population is seen as a driving force behind the change in focus from increasing GDP to increasing domestic consumption in the 12th Five Year Plan. [119] The national government still owns or controls many of the country's large industries and enterprises, and sets goals for economic development in the periodic Five Year Plans. As such, China's industries and enterprises are still encouraged to meet goals set in the national government's economic plan. But today's China also has a large and growing private sector, and improving the employment and income prospects for its citizens appears to be behind a shift to a market focus in many areas of China's "socialist market economy." [120]

The 12th Five Year Plan emphasizes the development new industries which can feed into the future growth of its economy. One such sector is renewable energy, and in the last five years, China has developed its manufacturing capabilities in wind turbines and solar panels and is using these capabilities to serve its own needs to produce renewable electricity. China recognizes that given the growing demand for energy at home, developing its domestic renewable energy industry and building manufacturing capacity can lead to advantages in future export markets. In late 2009, the Standing Committee of the National People's Congress affirmed its support of the development of the renewable power industry by building China's domestic capacity. Grid companies were directed to purchase all renewable electricity generated, with the State Council to determine the proportion of renewable electricity to overall generating capacity in order to meet national goals. [121] China has chosen to implement a Renewable Energy Law mandating minimum deployment levels for renewable electricity technologies (in terms of gigawatts of capacity by a target date), and to support these deployment goals with a complementary, national feed-in tariff for selected technologies.

Officially, however, energy efficiency and conservation remain China's top energy priority. These are considered the "low-hanging fruit" in the quest to reduce energy use and cut demand. Energy efficiency in China's larger firms is approaching levels in Western countries. The 11th FYP saw China's energy consumption per unit of GDP (i.e., energy intensity) drop by 19.6%.[122] The 12th FYP may not result in the energy savings seen in the previous five years, since many small power plants, less energy efficient coal-fired power plants, and steel mills were shut down. Few if any of these less efficient producers remain open.[123] Most of the energy savings in the 12th FYP are expected to come from structural changes in China's economic structure and technological improvements.[124] Energy efficiency benchmarks have been established for China's top 1,000 energy consuming industries (which accounted for 33% of national and 47% of industrial energy usage in 2004) with energy reduction goals set for each enterprise.[125] China is the world's largest market for new construction with approximately 2 billion square meters of floor space added annually. China's existing building codes (for both residential and public buildings) focus on heating, ventilation, and air conditioning, as well as lighting, hot water and power use. New standards have been in development since 2005 with national energy design criteria for residential buildings.[126] Energy efficient building codes are a key tool in establishing passive measures for energy savings in new future housing stock.

The weak link in China's electricity planning is transmission. Targets have been set for building renewable electric power generation without similar planning for accomplishing grid connection and integration, especially for remote wind power. China's grid companies must accept subsidized wind power, but costs of building transmission lines to connect to wind farms (especially those in remote regions) are not subsidized. As a result, as much as one-third of wind power generation in China is not connected to the grid due to a lack of transmission capacity. When combined with low tariffs, this likely means that China's wind power sector has been operating at a loss.[127] To help remedy the imbalance, the State Grid Corporation has begun to build a pilot "Smart Grid" in China's larger cities to help integrate renewable energy sources. Formal goals for a Smart Grid are expected to be incorporated into China's 12th Five Year Plan, with a "unified strong and smart grid" to be built nationwide by 2020.[128]

China today is dealing today with some of the same issues that the United States is likely to face as it considers building the infrastructure to take advantage of potentially huge wind and solar resources of the west and southwest. China's transmission system (like that of the United States) is mostly regional in functionality, and could benefit from improved connectivity across regions if renewable resources in remote areas are to be more fully harnessed. The cost of developing the transmission system will be great in both countries, but China has begun to build efficient ultrahigh voltage (UHV) transmission lines (i.e., voltages of 1,000 kilovolt (KV) alternating current, or higher, and 800 KV direct current, or higher). UHV can reduce transmission line losses and transmit more power over longer distances.[129] China's State Grid Corporation plans to invest $88 billion through 2020 to build UHV lines.[130]

In contrast to China's socialist market economy, the United States has largely a market-driven economy. Some observers argue that the current higher costs of renewable electricity do not favor market adoption.[131] However, the goals for increased use of renewable energy are diverse, and include energy security, energy independence, cleaner air, and more recently anthropogenic climate change, sustainability concepts, and economic development. Goals for reducing pollution have been enforced by government regulations requiring changes in the fuels used for power generation, and are often buttressed by requirements to install equipment to reduce particular types of emissions. Such goals could reasonably be said to apply both to the United States and China.

In the United States, while individual states may have renewable electricity mandates, there is no federal law creating a market for renewable energy. Many believe that renewable energy technologies need a federal policy driver which creates a national demand for renewable electricity if it is to be a "significant" contributor to domestic power generation picture. This opinion is largely based on the view that renewable energy technologies are not mature technologies, and therefore cannot yet compete with conventional fossil-fueled power generation. If increased deployment of renewable electricity technologies is a U.S. policy goal, a recent analysis by the National Renewable Energy Laboratory suggests that the United States could also implement both a renewable electricity standard and a FIT.[132] Others believe that renewable energy technologies should rely on venture capital and private sector investment alone, and the market alone should dictate whether they are employed. However, it is noted that the United States relies more on investment from venture capitalists for clean energy technology development than the rest of the world combined.[133] Venture capital has driven much energy innovation in the United States in the past and will undoubtedly play a role in the future in funding the next generations of clean energy technologies.

Creation of green jobs by developing a globally competitive renewable energy technology manufacturing industry has been described as a cornerstone of U.S. economic policy. Developing such a capability in the United States will likely require major investments from the federal government. Such an effort may have to go beyond tax credits and loan guarantees as have been used to this point in time if the development of renewable energy technology manufacturing capacity and "green jobs" creation are goals. In some countries, these policies can be instituted at the discretion of a central government. The massive allocation of government-funded resources by countries seeking to build renewable energy technology leadership and global market share is an indication of the greater reliance these nations expect to have on the renewable energy industry. The potential for product and equipment sales is key to their belief that such a transition should be accelerated.

The United States has traditionally relied primarily on market forces and tax incentives to encourage the development and deployment of new technologies. This would be the "business as usual" model. However, several forces are in play that call the "business as usual" model for innovation and deployment of renewable energy technologies into question: for example, investment dollars are scarce at this time, as the nation struggles to emerge from a recession; and other nations have aggressively used governmental powers to channel resources into renewable energy programs that have permitted them to establish renewable energy industries whose products and productivity often exceed those of the United States. Further, many believe that the U.S.'s existing (mostly non-renewable) low-cost energy system limits market opportunities in the short term, despite potential opportunities in the longer term or abroad. As a result, numerous observers and pundits have argued for more aggressive governmental intervention to bolster and accelerate U.S. activities relating to renewable energy. This is the backdrop for President Obama's declaration of a U.S. policy to invest in clean energy.

Like China, the United States relies most on coal for electric power generation with that fuel providing about half of the U.S. power generation.[134] Clean coal is seen as a part of China's clean energy future, and an opportunity exists for China and the United States to work together to develop clean coal technology.[135] However, China has decided that renewable electricity will also be a significant part of that energy future, with current plans for renewable energy sources to contribute 15% of its primary energy needs by 2020. Targets, planning, and investment have all followed to further China's renewable energy goals.

Unconventional sources of natural gas, such as coalbed methane and gas shales, may be resources that both the United States and China can develop.[136] Recent technological developments have raised the potential for natural gas to be produced in abundance especially from tight, impermeable shale formations. The outlook for renewable energy development could be affected if these unconventional natural gas sources can be developed and economically produced in an environmentally acceptable manner.[137]

End Notes

[1] Roger Ballentine, China Offers Tips on Using Energy More Efficiently, Renewable Energy World.com, February 23, 2009, http://www.renewableenergyworld.com/rea/news/article/2009/02/energy-efficiency-tips-from-china-54611.

[2] Energy Information Administration, International Energy Statistics, Total Electricity Installed Capacity , http://tonto.eia.doe.gov/cfapps/ipdbproject/IEDIndex3.cfm?tid=2&pid=2&aid=7.

[3] U.S. Energy Information Administration, China Energy Data and Statistics, July 2009, http://www.eia.doe.gov/ emeu/cabs/China/Background.html.

[4] About 60% of China's new coal plants have a fuel conversion to energy efficiency of about 44%, as compared to the most efficient coal plants in the United States with an efficiency of about 40%. Keith Bradsher, China Outpaces U.S. in Cleaner Coal-Fired Plants , New York Times, May 10, 2009, http://www.nytimes.com/2009/05/11/world/asia/ 11coal.html?_r=1.

[5] Greenhouse gases are carbon dioxide, methane, nitrous oxide, hydrofluorocarbons, perfluorocarbons, and sulfur hexafluoride.

[6] National Research Council, Hidden Costs of Energy: Unpriced Consequences of Energy Production and Use, October 19, 2009, http://www.nap.edu/catalog.php?record_id =12794#toc.

[7] Renewable energy resources are defined by the U.S. Department of Energy as energy resources that are naturally replenishing but flow-limited. They are virtually inexhaustible in duration but limited in the amount of energy that is available per unit of time. Renewable energy resources include biomass, hydro, geothermal, solar, wind, ocean thermal, wave action, and tidal action. See http://www.eia.doe.gov/glossary/glossary_r.htm.

[8] Jianxiang Yang, China Speeds Up Renewable Energy Development, Global Environmental Institute, October 26, 2006, http://www.worldwatch.org/node/4691.

[9] Xie Liangbing and Chen Yong , Making Up for Lost Time: China's Hydropower Push , Economic Observer, January 24, 2011, http://www.eeo.com.cn/ens/Industry/2011/ 01/24/192214.shtml. (CHHYDP).

[10] Xinhua News Agency, China to Significantly Raise Its Renewable Energy Targets, May 5, p. 2009, http://www.energy-enviro.fi/index.php?PAGE=2584.

[11] Official Chinese government (English language) source documents are used when possible. Amounts quoted in dollars use currency conversions in reference documents, and are not adjusted for time-value of money.

[12] The highest level body coordinating energy policy in China is the National Energy Commission established in January 2010. It is tasked with formulating energy development strategy, reviewing major issues related to energy security and energy development, and coordinating energy exploitation and international cooperation on energy issues. It is headed by Wen Jiabao, and its deputy director is China's Executive Vice Premier, Li Keqiang. Its members include the heads of ten government ministries, several quasi-ministries and regulatory commissions, the governor of the Central Bank, a deputy chief of the People's Liberation Army, and other power figures. China People's Daily online, China's National Energy Commission is established, January 27, 2010, http://english.peopledaily.com.cn/ 90001/90778/ 90862/6880658.html.

[13] "China bound its commitment domestically through a State Council decision even before last year's Copenhagen meeting, and it has said that the 40-45% carbon intensity reduction target will also be incorporated into its 12th Five Year Plan to be adopted by its National People's Congress in March 2011." See Deborah Seligsohn, Cancún Climate Change Summit: China's Journey from Copenhagen, Guardian Environmental Network, WRI, November 1, 2010, http://www.guardian.co.uk/environment/2010/dec/01/cancun-climate-change-summit-china.

[14] Pew Charitable Trusts, Who's Winning the Clean Energy Race?, March 2011, http://www.pewenvironment.org/uploadedFiles/PEG/Publications/Report/G-20Report-LO W Res-FINAL.pdf.

[15] Ministry of Science and Technology of the People's Republic of China, Key Technologies R&D Program, http://www.most.gov.cn/eng/programmes1/200610/t20061009_36224.htm.

[16] Center for American Progress (CAP), Out of the Running?, March 2010, http://www . americanprogress.org/issues/ 2010/03/pdf/out_ofjunning.pdf.

[17] CAP.

[18] CAP.

[19] The Five-Year Plan for National Economic and Social Development, or the Five-Year Plan, mainly aims to arrange national key construction projects, manage the distribution of productive forces and individual sector's contributions to the national economy, map the direction of future development, and set targets. See http://english.gov.cn/2006-04/05/content_245556.htm.

[20] HSBC Global Research, China's Next 5-year Plan—What It Means for Equity Markets, October 6, 2010, http://www.research.hsbc.com/midas/Res/RDV?p=pdf&key= DRpq 0Zsciy&n=279532.PDF.

[21] APCO Worldwide, China's 12th Five-Year Plan - How It Actually Works and What's in Store for the Next Five Years, December 10, 2010, http://www.apcoworldwide.com/content/PDFs/Chinas_12th_Five-Year_Plan.pdf.

[22] Department of Climate Change, National Development and Reform Commission, China Announces Targets for Carbon Emission, Energy Cuts for 2011, March 28, 2011, http://cdm.ccchina.gov.cn/english/NewsInfo.asp?NewsId= 5252.

[23] Angie Austin, Energy and Power in China: Domestic Regulation and Foreign Policy, The Foreign Policy Centre, April 2005, http://fpc.org.uk/fsblob/448.pdf.

[24] Article 48. See http://www.lehmanlaw.com/resource-centre/laws-and-regulations/ environ ment/electric-power-lawof-the-peoples-republic-of-china-1996.html.

[25] See http://www.china.org.cn/english/environment/34454.htm.

[26] China.org.cn, National Climate Change Program, http://www.china.org.cn/english /environ ment/213624.htm.

[27] Energy intensity as consumption per unit of gross domestic product.

[28] National Development and Reform Commission of China, National Climate Change Program, June 4, 2007, http://www.china.org.cn/english/environment/213624.htm.

[29] XinHua News, Stick to Energy Goals, May 7, 2010, http://news.xinhuanet.com/english2010/indepth/2010-05/07/ c_13281612.htm.

[30] Article 9. See http://www.china.org.cn/english/environment/34422.htm.

[31] Renewable Energy World, Authorized Release: The Renewable Energy Law. The People's Republic of China (herinafter RELaw), http://www.renewableenergyworld. com/assets/download/China_RE_Law_05.doc.

[32] http://www.chinaenvironmentallaw.com/wp-content/uploads/2008/04/medium-and-long-term-development-plan-forrenewable-energy.pdf.

[33] China Economic Net, China Adopts Amendment to Renewable Energy Law (herinafter (CENet), December 26, 2009, http://en.ce.cn/National/Politics/2009 12/26/t20091226_20695325.shtml.

[34] Peta Hodge, "China's Emissions Pledge Shows Commitment to Copenhagen," Green Business News, November 26, 2009, http://www.greenwisebusiness.co.uk/news/chinas-emissions-pledge-shows-commitment-to-copenhagen936.aspx.

[35] The National Energy Administration (NEA) was created in March 2008 as part of the general reforms of the Chinese energy sector by the Congress of the Communist Party of China. The NEA is a semiautonomous body under the NDRC responsible for energy planning and development, drafting of energy laws, and international energy cooperation. See http://ec.europa.eu/energy/international/bilateral_cooperation/china/stakeholders_en.htm.

[36] Barbara Finamore, Staff Blog - China's Announcements on Energy and Climate in Advance of Presidential Summit, Natural Resources Defense Council, January 18, 2011, http://switchboard.nrdc.org/blogs/bfinamore/ chinas_announcements_on_energy.html.

[37] David DuByne, Biogas? China size it, Science Alert, May 9, 2008, http://www. sciencealert. com.au/opinions/ 20080905-17301.html.

[38] http://www.cite-sciences.fr/france-chine/en/chinaworld/2/c15.html.

[39] Methane digesters convert manure or other organic matter into biogas through a process called anaerobic digestion. In this process, bacteria decompose the organic matter in the absence of oxygen, producing a gas composed of 60 to 70 percent methane and 30 to 40 percent carbon dioxide—biogas. See http://www.cleanenergyresourceteams.org/ technology/biogas-digesters.

[40] By the end of 2005, there were more than 17 million household biogas digesters producing 6,500 million cubic meters of biogas annually. Over 1,500 large-and medium-scale digester projects facilities generate around 1500 million cubic meters of biogas annually. See http://www.china.org.cn/english/environment/213624.htm.

[41] Renewable Energy Policy Network for the 21st Century, Recommendations for Improving the Effectiveness of Renewable Energy Policies in China, October 2009, http://www.ren21.net/pdf/ Recommendations_for_RE_Policies_in_China.pdf. (RECREN).

[42] A feed-in tariff is an energy-supply policy focused on supporting the development of new renewable power generation. ...The FIT contract provides a guarantee of payments in dollars per kilowatt hour for the full output of the [renewable energy] system for a guaranteed period of time (typically 15-20 years). Karlynn Cory, Toby Couture, and Claire Kreycik, Feed-in Tariff Policy: Design, Implementation, and RPS Policy Interactions, National Renewable Energy Laboratory, NREL/TP-6A2-45549, March 2009. (FITP).

[43] RECREN.

[44] RECREN.

[45] Zhang Qi, Burning Willows to Stop Desertification, China Daily, May 11, 2009. http://www.chinadaily.com.cn/bw/ 2009-05/11/content_7761535.htm.

[46] The potential land area for cultivating oilseed plants and energy crops (including jatropha curcas, rapeseed, ricinus communis, lacquer tree, Chinese goldthread tree, and sweet sorghum) is estimated to meet the annual feedstock requirements of 50 million tons of liquid biofuel. China has banned the use of grain for ethanol production. RECREN.

[47] Jonathan Lynn, China Became Net Food Importer in 1st Half, Reuters, August 22, 2008, http://www.planetark.com/ dailynewsstory.cfm/newsid/49900/story.htm.

[48] Tim Weckesser, The China Business Network, Number 1 in Thermal, Solar, Nuclear, Wind and Hydro Power, January 27, 2011, http://www.thechinabusinessnetwork.com/Channel-Index/Green-Development/General/General2011/Number-1-in-Thermal,-Solar,-Nuclear,-Wind-and-Hydr.aspx.

[49] energychinaforum.com, NEA Outlines Key Tasks for Energy Sectors, to Benefit Related Listed Firms, January 20, 2011, http://www.energychinaforum.com/news/46032.shtml.

[50] National Development and Reform Commission, Medium and Long-Term Development Plan for Renewable Energy in China, September 2007, http://www.cresp.org.cn/uploadfiles/ 2/967/medium%20and%20long-term%20development%20plan%20for%20re%20in%20china%20eng.pdf.

[51] See CHHYDP.

[52] Shanghai Daily, China ready for flood of hydropower, January 6, 2011, http://www.phayul. com/news/article.aspx? id=28883.

[53] Peter Bosshard, Dam Nation, ForeignPolicy.com, March 8, 2011, http://www. foreignpolicy.com/articles/2011/03/ 08/dam_nation.

[54] Whats On Xiamen, Inc., China Trina Solar to Invest $US800m in New Manufacturing Plants, December 30, 2010, http://www.whatsonxiamen.com/tech593.html.

[55] Energychinaforum.com, India and China to Benefit as Global Solar PV Returns to Pre-Recession Growth, March 8, 2011, http://www.energychinaforum.com/news/47839.shtml.

[56] http://earth2tech.com/2009/06/24/is-china-on-the-cusp-of-becoming-a-huge-solar-panel-market/

[57] Breakthrough Institute and the Information Technology and Innovation Foundation, Rising Tigers, Sleeping Giant (herinafter Rising Tigers), November 2009, http://thebreakthrough.org/blog/Rising_Tigers.pdf.

[58] China also announced several utility-scale renewable energy projects in 2009, including the world's largest wind farm, a 10 gw "Three Gorges of Wind Power" project in Gansu Province, and a 2 gw solar power plant in Northern China using Arizona-based First Solar's thin-film solar PV panels. See http://switchboard.nrdc.org/blogs/bfinamore/ china_ records_ its_climate_acti.html.

[59] Yotam Ariel, Incentives, Falling Cost and Rising Demand in China's PV Market, Renewable Energy World, November 13, 2009. http://www.renewableenergyworld. com/rea/ news/article/2009/11/incentives-falling-cost-andrising-demand-in-chinas-pv-market.

[60] Chris Brown, NDRC's Zhang Guobao and the Feed-in Tariff, China Solar Energy, April 7, 2010, http://chinasolarenergy.blogspot.com/2010/04/ndrcs-zhang-guobao-and-feed-in-tariff.html.

[61] Helios CSP, China Moves Toward Concentrating Solar Power Plants, February 23, 2011, http://helioscsp.com/ noticia.php?id_not=221.

[62] Whats On Xiamen, Inc., China Overtakes US as World's Largest Wind-Power Installer, January 14, 2011, http://www.whatsonxiamen.com/tech625.html.

[63] Competitive bids for concessions with a goal of reducing the price of wind power projects. See http://www.gwec.net/ index.php?id=125.

[64] With the exception of Guandong Province which approves its own projects and prices. RECREN.

[65] Ecoworld, Wind Power in China, July 2006, http://www.ecoworld.com/energy-fuels/wind-power-in-china.html.

[66] RECREN.

[67] RELaw.

[68] Tariff levels ranged from $0.07 to $0.09 per kwh. Rising Tigers.

[69] Zhang Qi, Wind Can Power Up Entire Nation, China Daily News, June 18, 2009, http://www.chinadaily.com.cn/ bizchina/2009-06/18/content_8296706.htm.

[70] New York Times, "Gains in Global Wind Capacity Reported," February 15, 2010, http://green.blogs.nytimes.com/ 2010/02/15/gains-in-global-wind-capacity-reported/.

[71] China to Triple Wind Goal to 100,000 mw by 2020, May 6, 2009, http://climateprogress.org/2009/05/06/energy-andglobal-warming-news-china-solar-wind/.

[72] China.Org.Cn, Xinhua, China Speeds Up Offshore Wind Power Construction, March 20, 2010, http://www.china.org.cn/business/2010-03/20/content_19647071.htm.

[73] National Renewable Energy Laboratory, Renewable Energy in China—Financial Incentives, NREL/FS-710-36045, April 2004, http://www.nrel.gov/docs/fy04osti/36045.pdf.

[74] The original surcharge of 0.014 cents per kwh was increased to 0.029 cents per kwh in 2007. See RECREN, p. 23.

[75] For example, electricity for residential use in Beijing was 49 fen per kwh [about $0.07], while that for agricultural use was around 52 fen per kwh, for secondary industry use 76 fen per kwh and for commercial use 79 fen per kwh, according to Beijing Electric Power Corporation. One yuan is 100 fen. See http://news.xinhuanet.com/english/2009-11/ 19/content_12492364.htm.

[76] RECREN.

[77] Keith Bradsher, China Leading Global Race to Make Clean Energy, New York Times, January 30, 2010 (herinafter NYT1), http://www.nytimes.com/2010/01/31/business/energy-environment/31renew.html.

[78] NYT1.

[79] Article 10. See http://english.gov.cn/laws/2005-10/08/content_75023.htm.

[80] See http://www.chinadaily.com.cn/bizchina/2009-11/24/content_9033778_3.htm.

[81] See China's Promotion of the Renewable Electric Power Equipment Industry (herinafter CPRE), http://www.nftc.org/default/Press%20Release/2010/China%20 Renewable%20 Energy.pdf.

[82] The national government established the National Bio Energy Company, Limited, as a joint venture between the State Grid Corporation of China and the Dragon Power Company, Ltd. CPRE.

[83] Xinhua, China to Offer Incentives for Non-Food Biofuels, November 7, 2007, http://www. chinadaily.com.cn/china/ 2007-12/07/content_6306076.htm.

[84] See Global Subsidies Initiative of the International Institute for Sustainable Development, BIOFUELS—AT WHAT COST ? Government support for ethanol and biodiesel in China, December 2008, http://www.globalsubsidies.org/files/ assets/China_Biofuels_Subsidies.pdf.

[85] Zhu Shaobin, China's Economic Stimulus Plans Benefit Environment, Xinhua, March 3, 2009, http://news.xinhuanet.com/english/2009-03/10/content_10986048.htm.

[86] Sewell Chan and Keith Bradsher, "U.S. to Investigate China's Clean Energy Aid," New York Times, October 15, 2010, http://www.nytimes.com/2010/10/16/business/ 16wind.html?_r =1&pagewanted=print.

[87] A group of 17 elements consisting of scandium, yttrium, and the 15 lanthanides. Rare-earth elements are vital to many electronic and renewable energy technologies. See CRS Report R41347, Rare Earth Elements: The Global Supply Chain, by Marc Humphries.

[88] Robert F. Service, Chinese Policies Could Pinch U.S. Efforts to Make Electric Vehicles, American Association for the Advancement of Science, Science, Vol. 329 no. 5990 p. 377, July 23, 2010, http://www.sciencemag.org/content/ 329/5990/377.1.short.

[89] John Miller and James Areddy, Trade Judges See Flaw in China Policies: Preliminary WTO Report Finds No Case for Some of Beijing's Export Restrictions, Wall Street Journal, February 18, 2011.

[90] Michael Wines and Xiyun Yang, "China Rips US Complaint on Clean Energy Aid," New York Times, October 18, 2010, http://www.boston.com/news/world/asia/articles/ 2010/10/18/china_rips_us_complaint_on_clean_energy_aid/.

[91] CRS Report R40155, Renewable Fuel Standard (RFS): Overview and Issues, by Randy Schnepf and Brent D. Yacobucci

[92] The Union of Concerned Scientists projects that state Renewable Portfolio Standard programs will support 76,750 mw of new renewable power by 2025—an increase of 570% over total 1997 U.S. levels (excluding hydro). http://www.ucsusa.org/assets/documents/clean_ energy/RES_in_the_States_Update.pdf

[93] http://www.ferc.gov/market-oversight/othr-mkts/renew/othr-rnw-rps.pdf

[94] See business energy tax credit discussion at http://www.dsireusa.org/incentives/incentive. cfm?Incentive_Code= US02F&re=1&ee=1.

[95] National Academy of Sciences, Energy Research at DOE: Was It Worth It? Energy Efficiency and Fossil Energy Research 1978 to 2000, Washington, DC, 2001, http://www.nap. edu/catalog.php?record_id=10165.

[96] Qualifying Facility (QF): A cogeneration or small power production facility that meets certain ownership, operating, and efficiency criteria established by the Federal Energy Regulatory Commission (FERC) pursuant to the Public Utility Regulatory Policies Act (PURPA). See http://www.eia.doe.gov/glossary/glossary_q.htm.

[97] Under the Energy and Security Act of 1980 were the following: U.S. Synthetic Fuels Corporation Act, Biomass Energy and Alcohol Fuels Act, Renewable Energy Resources Act, Solar Energy and Energy Conservation Act and Solar Energy and Energy Conservation Bank Act, Geothermal Energy Act, and the Ocean Thermal Energy Conversion Act.

[98] See http://www1.eere.energy.gov/femp/regulations/epact1992.html#wc.

[99] The renewable electricity Production Tax Credit is a per-kilowatt-hour tax credit for electricity generated by qualified renewable energy technologies and sold by the taxpayer to an unrelated party during the taxable year. See CRS Report R40913, Renewable Energy and Energy Efficiency Incentives: A Summary of Federal Programs, by Lynn J. Cunningham and Beth A. Roberts.

[100] CRS Report R40412, Energy Provisions in the American Recovery and Reinvestment Act of 2009 (P.L. 111-5), coordinated by Fred Sissine.

[101] CRS Report R40833, Renewable Energy—A Pathway to Green Jobs?, by Richard J. Campbell and Linda Levine.

[102] U.S. Energy Information Administration, Annual Energy Outlook 2011 Early Release Overview (herinafter AEO 2011), Renewable Energy Generating Capacity & Generation, March 2011, http://www.eia.gov/oiaf/aeo/tablebrowser/ #release=AEO2011&subject=0-AEO2011&table=16-AEO2011&region=0-0&cases=ref2011-d120810c.

[103] Ibid.

[104] CRS Report R41282, Agriculture-Based Biofuels: Overview and Emerging Issues, by Randy Schnepf.

[105] CRS Report RL34738, Cellulosic Biofuels: Analysis of Policy Issues for Congress, by Kelsi Bracmort et al.

[106] See http://www.nrel.gov/biomass/pdfs/40742.pdf.

[107] Oak Ridge National Laboratory, Dams: Multiple Uses and Types, http://www.ornl.gov/info/ornlreview/rev26-34/ text/hydside1.html.

[108] U.S. Department of Energy - Idaho National Laboratory, Feasibility Assessment of the Water Energy Resources of the United States for New Low Power and Small Hydro Classes of Hydroelectric Plants, DOE-ID-11263, January 2006, http://hydropower.inel.gov/resourceassessment/pdfs/main_report_appendix_a_final.pdf.

[109] AEO 2011.

[110] AEO 2011.

[111] CRS Report R40833, Renewable Energy—A Pathway to Green Jobs?, by Richard J. Campbell and Linda Levine..

[112] See http://www.windpoweringamerica.gov/pdfs/2007_annual_wind_market_report.pdf.

[113] Reuters, Cape Wind, first U.S. offshore wind farm, approved, April 28, 2010, http://www.reuters.com/article/ idUSTRE63R42X20100428.

[114] Walt Musial, Offshore Wind Energy Potential for the United States, National Renewable Energy Laboratory, May 2005. http://www.windpoweringamerica.gov/pdfs/ workshops/2005_summit/musial.pdf.

[115] CRS Report R41040, Identifying Incentives and Barriers to Federal Agencies Achieving Energy Efficiency and Greenhouse Gas Reduction Targets, by Anthony Andrews and Richard J. Campbell.

[116] U.S. Department of Energy, Office of Energy Efficiency and Renewable Energy, Renewable Energy Production Incentive, April 10, 2007, http://apps1.eere.energy.gov/repi/about.cfm.

[117] The current payment under REPI is $0.021 per kwh. If there are insufficient appropriations to make full payments for electricity production from all qualified systems for a federal fiscal year, 60% of the appropriated funds for the fiscal year will be assigned to facilities that use solar, wind, ocean, geothermal or closed-loop biomass technologies; and 40% of the appropriated funds for the fiscal year will be assigned to other eligible projects. Funds will be awarded on a pro rata basis, if necessary. See http://www.dsireusa.org/incentives/incentive.cfm?Incentive_Code=US33F&re= 1&ee=1.

[118] See http://www.dsireusa.org/incentives/incentive.cfm?Incentive_ Code=US05F&State=Federal&currentpageid=1.

[119] S. Klamp, Chinese 12th 5-year plan – New Energy, New Energy Cars, Wordpress.com, November 8, 2010, http://cleaninvest.wordpress.com/2010/11/08/chinese-12th-5-year-plan-new-energy-new-energy-cars/.

[120] CRS Report RL30519, The Growth of the Private Sector in China and Implications For China's Accession to the World Trade Organization, by Wayne M. Morrison.

[121] CENet.

[122] Liu Jianqiang, Reining in China's Energy Targets, Chinadialogue, March 2, 2011, http://www.chinadialogue.net/article/show/single/en/4138-Reining-in-China-s-energy-targets.

[123] Reuters, "China Small Power Plants Closures Ahead of Schedule," July 3, 2010, http://in.reuters.com/article/2010/07/03/idINIndia-49857620100703.

[124] Yang Fuqiang, Hu Yangli, and Li Jingjjng, China Needs Higher Targets, Chinadialogue.net, March 3, 2011, http://www.chinadialogue.net/article/show/single/en/4140—China-needs-higher-targets-.

[125] Lynn Price, Xuejun Wang, and Jiang Yun, China's Top-1000 Energy-Consuming Enterprises Program: Reducing Energy Consumption of the 1000 Largest Industrial Enterprises in China, Ernest Orlando Lawrence Berkeley National Laboratory, LBNL-519E, June 2008.

[126] B. Shui, M. Evans, and H. Lin, et al., Country Report on Building Energy Codes in China, Pacific Northwest National Laboratory, April 2009, http://www.energycodes.gov/implement/pdfs/CountryReport_China.pdf.

[127] China's wind power sector is expected to be operating at a loss for some time to come, largely due to the practice of awarding concessions to the bids with the lowest tariff and awarding contracts for the life of a project. As such, higher government subsidies may be required for wind farm profitability, and building transmission infrastructure will still be needed on a massive scale to bring this power to market. See Guardian.co.uk, "Wind Power Growth in China's Deserts Ignored Financial Risks," May 14, 2010, http://www.guardian.co.uk/environment/2010/may/14/wind-power-chinadesert.

[128] The "Smart Grid" is an intelligent system capable of seamlessly integrating renewable energy sources like solar and wind power into the electricity network. See Xinhua News, "China Sets to Build Smart Grid to Tap Renewable Energy," March 13, 2010, http://news.xinhuanet.com/english2010/china/2010-03/13/c_13209617.htm.

[129] International Electrotechnical Commission, Energy Efficient Ultra High Voltage: The Future of Electricity Transmission, March 2007, http://electronics.ihs.com/news/articles/iec-uhv-electricity-transmission.htm.

[130] Julian Wong and Andrew Light, China Begins Its Transition to a Clean-Energy Economy, Center for American Progress, June 4, 2009, http://www.americanprogress. org/issues/2009/06/china_energy_numbers.html.

[131] Jerry Taylor and Peter Van Doren , Evaluating the Case for Renewable Energy: Is Government Support Warranted?, CATO Institute, Policy Analysis no. 422, January 10, 2002, http://www.cato.org/pub_display.php? pub_id=1281.

[132] FITP.

[133] "The United States continues to hold an overwhelming advantage in the area of venture capital/private equity investment, accounting for 73 percent of the G-20 total in 2010." See Pew Charitable Trusts, Who's Winning the Clean Energy Race?, March 2011, http://www.pewenvironment.org/uploadedFiles/PEG/Publications/Report/G-20Report-LOWRes-FINAL.pdf.

[134] U.S. Energy Information Administration, Net Generation by Energy Source: Total (All Sectors), May 14, 2010, http://www.eia.doe.gov/cneaf/electricity/epm/table1_1.html.

[135] See http://www.energy.gov/news2009/documents2009/US-China_Fact_Sheet_Coal.pdf.

[136] Vello Kuuskraa and Scott Stevens, Worldwide Gas Shales and Unconventional Gas—A Status Report, Advanced Resources International, December 12, 2009, http://www.adv-res.com/pdf/ Kuuskraa%20Condensed%20Worldwide%20Uncon%20Gas%2012_12_09.pdf.

[137] CRS Report R40894, Unconventional Gas Shales: Development, Technology, and Policy Issues, coordinated by Anthony Andrews.

In: China and the U.S.
Editor: Julian A. Paquette
ISBN: 978-1-62948-389-4

Chapter 2

A COMPARISON OF IRON AND STEEL PRODUCTION ENERGY USE AND ENERGY INTENSITY IN CHINA AND THE U.S.*

Ali Hasanbeigi, Lynn Price, Nathaniel Aden, Zhang Chunxia, Li Xiuping and Shangguan Fangqin

ABSTRACT

Production of iron and steel is an energy-intensive manufacturing process. In 2006, the iron and steel industry accounted for 13.6% and 1.4% of primary energy consumption in China and the U.S., respectively (U.S. DOE/EIA, 2010a; Zhang et al., 2010). The energy efficiency of steel production has a direct impact on overall energy consumption and related carbon dioxide (CO_2) emissions. The goal of this study is to develop a methodology for making an accurate comparison of the energy intensity (energy use per unit of steel produced) of steel production. The methodology is applied to the steel industry in China and the U.S. The methodology addresses issues related to boundary definitions, conversion factors, and indicators in order to develop a common framework for comparing steel industry energy use.

This study uses a bottom-up, physical-based method to compare the energy intensity of China and U.S. crude steel production in 2006. This year was chosen in order to maximize the availability of comparable

* This is an edited, reformatted and augmented version of Lawrence Berkeley National Laboratory, Publication No. LBNL-4836E, dated June 2011.

steel-sector data. However, data published in China and the U.S. are not always consistent in terms of analytical scope, conversion factors, and information on adoption of energy-saving technologies. This study is primarily based on published annual data from the China Iron & Steel Association and National Bureau of Statistics in China and the Energy Information Agency in the U.S.

This report found that the energy intensity of steel production is lower in the United States than China primarily due to structural differences in the steel industry in these two countries. In order to understand the differences in energy intensity of steel production in both countries, this report identified key determinants of sector energy use in both countries. Five determinants analyzed in this report include: share of electric arc furnaces in total steel production, sector penetration of energy-efficiency technologies, scale of production equipment, fuel shares in the iron and steel industry, and final steel product mix in both countries. The share of lower energy intensity electric arc furnace production in each country was a key determinant of total steel sector energy efficiency. Overall steel sector structure, in terms of average plant vintage and production capacity, is also an important variable though data were not available to quantify this in a scenario. The methodology developed in this report, along with the accompanying quantitative and qualitative analyses, provides a foundation for comparative international assessment of steel sector energy intensity.

1. Introduction

Production of iron and steel is an energy-intensive manufacturing process. In 2006, the iron and steel industry accounted for 13.6% and 1.4% of primary energy consumption in China and the U.S., respectively (U.S. DOE/EIA, 2010a; Zhang et al., 2010).[1] The energy efficiency of steel production has a direct impact on overall energy consumption and related carbon dioxide (CO_2) emissions. The goal of this study is to develop a methodology for making an accurate comparison of the energy intensity of steel production in China and the U.S. Energy intensity is defined as energy use per unit of steel produced. Energy efficiency is defined as the ratio of useful output of service/product from a piece of industrial equipment or an industrial process to the energy used by that equipment/process. The methodology addresses issues related to boundary definitions and conversion factors in order to develop a common framework for comparing steel industry energy use in these two countries.

The areas covered in this study are: (1) a literature review of existing research on energy efficiency and intensity of steel production in China and the U.S. and comparisons of energy intensity levels between the two countries, (2) a review of the available data on technology, production, and energy use for the Chinese and U.S. steel industries, (3) development of a methodology with consistent boundaries and conversion factors for comparing energy intensity of steel production in China and the U.S., (4) comparison of the energy intensity of steel production in China and the United States, and (5) development of explanatory variables to explain the differences in energy intensity.

This report begins with an overview of the iron and steel industry in China and the U.S. in Section 2. Section 3 describes the methodology used for this study, including an assessment of the data available in both countries, definition of boundaries of the iron and steel industry used in this study, and issues related to the calculation of the energy content of fuels as well as conversion factors to convert electricity from final to primary energy[2] for use in this study. Section 4 outlines the base year production and energy use data for both countries. Section 5 presents the quantitative comparison of energy intensity of iron and steel production in China and the U.S. This comparison provides a base case and six alternative scenarios that present the results using different assumptions than those used in the base case. Explanatory variables such as the share of electric arc furnace (EAF) steel production, product mix, fuel shares, etc. are also qualitatively discussed in Section 5. Section 6 summarizes study findings, while Section 7 provides a discussion, policy implications and recommendations for future research.

2. Overview of the Iron and Steel Industry in China and the U.S.

2.1. The Iron and Steel Industry in China

China is a developing country and is currently in the process of industrialization. The iron and steel industry, as a pillar industry for Chinese economic development, has grown rapidly along with the national economy. Starting in the 1990s, the industry development accelerated, with crude steel production in 1996 exceeding more than 100 million metric tonnes (Mt). Since then, steel production in China has continued to increase rapidly, and China has been the world's largest crude steel producer for 14 continuous years. The average annual growth rate of crude steel production was 18.5% between 2000 and 2009. Steel production in 2010 was 627 Mt (worldsteel, 2011), representing 46.6% of the world production that year (see Figure 1).

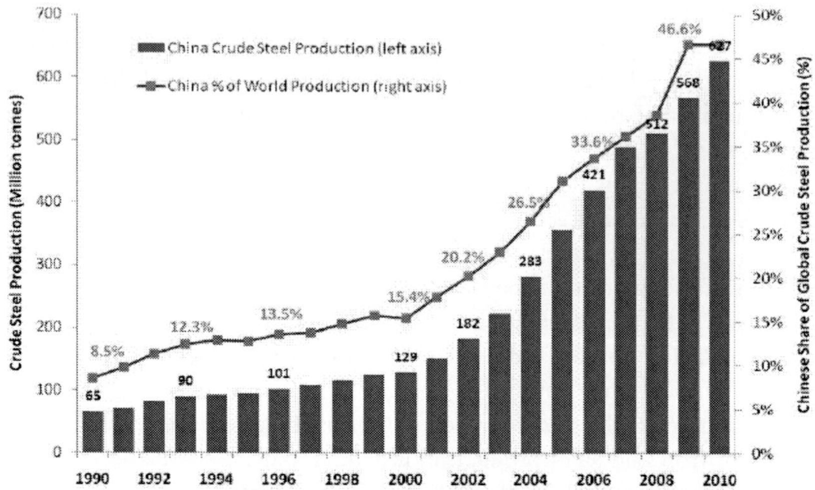

Source: China Iron and Steel Industry Yearbook, various years; World Steel Association 2011.

Figure 1. China's Crude Steel Production and Share of Global Production (1990-2010).

The Chinese iron and steel industry has made much progress in reducing energy use, starting from energy saving on individual equipment and process energy conservation in 1980s to systematic energy conservation via process optimization in 1990. China's energy consumption per tonne of steel has declined significantly, especially since the 1990s, largely due to process restructuring and optimization (see Table 1). Energy intensity in China is measured in units of kilograms of coal equivalent per metric tonne (kgce/tonne).[3] During the ten years between 1990 and 2000, China's steel production almost doubled, but total energy consumption only increased 31%. From 2000 to 2005, steel production increased 174.2%, but energy consumption only increased 120% (Editorial Board of China Iron and Steel Industry Yearbook, various years). Specific energy consumption per tonne of steel in *key medium and large - sized steel enterprises*[4] dropped from 920 kgce/t steel in 2000 to 741 kgce/t steel in 2005 (Editorial Board of China Iron and Steel Industry Yearbook, various years). [5] Specific energy consumption per tonne of steel was reduced 19.5% from 2000 to 2005. Table 2 provides more detailed information on the reduction of energy intensity in the main processes of key steel enterprises.

Table 1. Energy Savings and Components of China's Steel Industry (1990-1999)

Component Type	Energy savings (Mtce)	Energy savings (%)
Optimization of steel production process a	18.7	40.7%
Use of energy-saving technologies and equipment	8.71	18.9%
Steel production auxiliary materials improvements and others	7.07	15.4%
Energy management	11.5	25.0%
Total	46.0	100%

Source: Yin RY, et al., 2003.

[a] This category refers to upgrading the steel making process to be more efficient and more environmentally-friendly (e.g., by using more alternative fuels and wastes). Specific measures may include: upgrading technologies and equipment, upgrading the steel product structure (by producing more value-added steel products), and increasing the quality of steel products. The goal is to increase the energy efficiency of steel making and to utilize more wastes (scrap, plastics, municipal wastes, tires, sewage sludge, etc.).

Since 2000, energy conservation and emission reduction in China's steel industry has improved significantly (see Table 2 for *key medium and large - sized steel enterprises).* Academic advisors recommended that the steel industry explore the functions of steel product manufacturing, energy conversion, and utilization and treatment of waste resources (Yin, 2009). This focus leads to energy conservation and emission reduction in the steel industry. Meanwhile, China's national government is actively promoting the concept of a circular (or recycling) economy in the steel industry, encouraging widespread energy saving, emission reduction, increased steel scrap recycling rate, and resource conservation as necessary foundations of the circular economy. In addition, energy conservation is also seen as an effective way of reducing greenhouse gas emissions.

Under the guidance of the concept of "expanding the functions of steel manufacturing processes," promotion and application of energy-saving technologies has already become an important step for increasing energy efficiency and reducing energy consumption of steel enterprises. During this time, energy-conservation technologies adopted in China include: Coke Dry

Quenching (CDQ), Top-pressure Recovery Turbine (TRT), recycling converter gas, recycling waste heat from converter steam, continuous casting, slab hot charging and hot delivery, Coal Moisture Control (CMC), and recycling waste heat from sintering. The penetration level of energy-conservation technologies in the steel industry has improved greatly in China, increasing energy conservation and emission reductions.

Table 2. Changes in Energy Intensity of Key Medium and Large-sized Chinese Steel Enterprises and in the Main Steel-Making Processes (2000-2008)

Year	Comprehensive Energy Consumption per tonne of steel (kgce/t)	Energy Intensity of Main Processes (kgce/t)				
		Coking	Sintering	Iron-making	Basic Oxygen Furnace (BOF)	Electric Arc Furnace (EAF)
2000	920	160.20	68.90	466.07	28.88	265.59
2001	876	153.98	68.60	452.01	28.03	230.09
2002	815	150.32	67.07	455.13	24.01	228.94
2003	770	148.51	66.42	464.68	23.56	213.73
2004	761	142.21	66.38	466.20	26.57	209.89
2005	741	142.21	64.83	456.79	36.34	201.02
2006	645	123.11	55.61	433.08	9.09	/
2007	628	121.72	55.21	426.84	6.03	/
2008	630	119.97	55.49	427.72	5.74	/

Source: Editorial Board of China Iron and Steel Industry Yearbook, various years.

Notes:

(1) Data in the table are from member companies of the China Iron and Steel Association.

(2) In the reported statistics, a primary energy conversion factor of 0.404 kgce/kWh was used for electricity during 1900-2005; and final energy conversion factor of 0.1228 kgce/kWh was used for electricity during 2006-2008. This is the primary reason for the large difference between the 2005 and 2006 data.

(3) Since 2006, the refining process of the BOF energy consumption is calculated separately.

(4) To convert units from kgce/t to GJ/t, multiply the values by 0.02931.

2.2. The Iron and Steel Industry in the U.S.

In the U.S., steel production peaked in 1973 at a level of 137 Mt (USGS, 2010a). As total production declined, the U.S. steel industry phased out open hearth furnaces and increased the share of electric arc furnace production from 18% in 1973 to 57% in 2006, growing to 64% in 2009 (worldsteel, 2009). Figure 2 shows total U.S. steel production by technology type. After 2000, the level of U.S. steel production hovered below 100 Mt, with total production of 98 Mt in 2006. U.S. steel production in 2009 dropped to 56 Mt - a 19 Mt drop in crude steel production in one year (USGS, 2010b), but rebound to about 80 Mt in 2010 (worldsteel, 2011).[6]

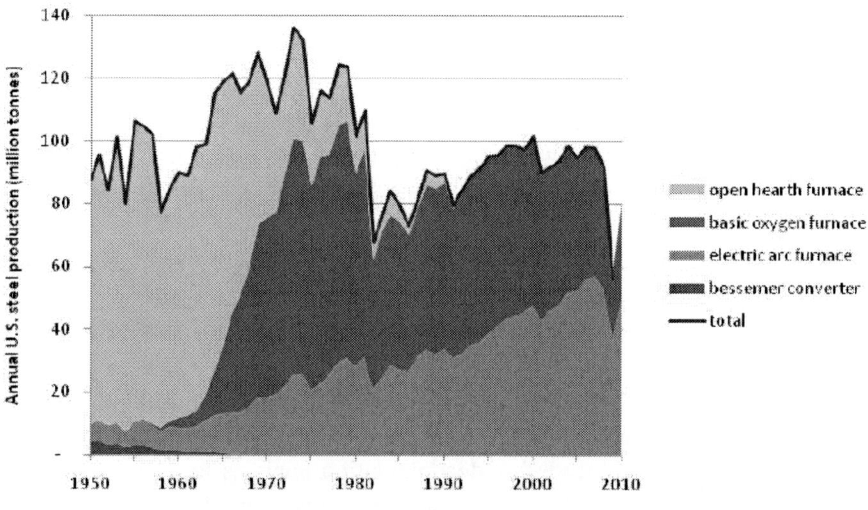

Source: AISI, 2008; USGS, 2010a; worldsteel, 2011.

Figure 2. U.S. Crude Steel Production by Technology Type (1950-2010).

The energy intensity and energy efficiency of American steel production continuously improved due to industry restructuring in the 1970s and 1980s, an increase in production of steel in electric arc furnaces, adoption of continuous casting, use of direct hot rolling, and feedstock process improvements (Ruth, 2000; Tornell, 1997).

In 2000, an analysis found that the energy intensity of steel production in the U.S. dropped from 20 to 18 MBtu/t (23.3 to 20.9 GJ/tonne, respectively) between 1990 and 1998 (Stubbles, 2000). The American Iron and Steel

Institute (AISI) reported that the energy intensity of U.S. shipped steel production improved by 15% between 2002 and 2006 (AISI, 2008).

2.3. Previous Comparisons of China and U.S. Steel Industry Energy Use and Intensity

Previous comparisons of international steel industry energy use and intensity have employed a range of methods. Worrell et al. (1997) found that physical-based indicators provided a more robust basis for international comparison than economic indicators of steel sector energy efficiency and intensity. Within the range of physical-based analyses, a variety of study boundaries, units of analysis, and conversion factors are used. For example, whereas Worrell et al. (1997) use crude steel production as their unit of analysis, Stubbles (2000) calculated energy use and emissions per ton of shipped steel. Likewise, whereas Andersen and Hyman (2001) include coke-making energy use, Kim and Worrell (2002) omit coke making.

International comparisons of steel production energy efficiency and CO_2 emissions were published in the early 2000s with data through the 1990s. Worrell et al. (1997) used physical production index (PPI) decomposition analysis to identify process and product mix factors in overall energy intensity changes. Over the period from 1980 to 1991, Worrell et al. found the specific energy consumption (energy use per tonne of steel) of iron and steel production to be approximately 60% higher in China than the United States.

This comparative research was updated in 2002 when Kim and Worrell (2002) used CO2 intensity index decomposition analysis to examine the CO2 intensity of steel production in seven countries, including China and the United States. The analysis found that increased activity levels, structural demand shifts to a more energy-intensive steel product mix, and increased use of coal counterbalanced process efficiency improvements, thereby slowing the reduction of Chinese steel production CO2 intensity between 1980 and 1996. In their U.S. analysis, Kim and Worrell (2002) found that process efficiency improvements were the most important factor for explaining the reduced overall CO2 intensity of steel production. The carbon intensity given for China in 1994 is 0.85 tC/ton and for the U.S. is 0.55 tC/ton which is 35% lower than that on in China (Kim and Worrell 2000). The energy intensity and CO_2 intensity differences between two countries are not quite the same, as CO_2 intensity depends also to the grid CO_2 emission factor, EAF ratio, etc.

The International Energy Agency (IEA) also includes country-specific steel sector energy analysis in their overview of potential energy efficiency savings to 2050 (IEA, 2010). Based on 2007 data, the IEA found that China could save 6.1 GJ/tonne crude steel and the U.S. could save 2.4 GJ/tonne of crude steel through adoption of best available technologies. While the IEA does not present an explicit comparison of steel production in China and the U.S., its analysis implies that U.S. steel production was on average 3.7 GJ/tonne crude steel less energy intensive than Chinese production in 2007.

A review of the above comparison studies found that boundary and conversion factor assumptions are not always explicitly stated, but appear to vary widely, especially for characterizing imported or off-site produced inputs. Published studies indicate that consensus has yet to form on boundaries and conversion factors for international comparison of steel production energy efficiency, resulting in widely disparate results which are difficult to interpret and compare.

3. METHODOLOGY

This study uses a bottom-up, physical-based methodology to compare the energy intensity of China and U.S. crude steel production in 2006. This year was chosen in order to maximize the availability of comparable steel-sector data. However, data published in China and the U.S. are not always consistent in terms of analytical scope, conversion factors, and information on adoption of energy-saving technologies. This study is primarily based on published annual data from the China Iron and Steel Association and National Bureau of Statistics in China and the Energy Information Administration in the U.S.

3.1. Data Availability and Base Year for Analysis

3.1.1. Data Availability - China

There are two meaningful sources of total energy consumption data for China's iron and steel industry: one is from former Ministry of Metallurgical Industry (or current China Iron and Steel Association); and the other one is National Bureau of Statistics (see Table **3**). The *China Steel Industry Yearbook* and *China Steel Statistics* are managed by the China Iron and Steel Association, and only provide national energy consumption data up to 2003. The *China Energy Statistical Yearbook* is published by the Chinese National

Bureau of Statistics, and has more comprehensive data for the steel industry's energy consumption. However, because the *China Energy Statistical Yearbook* only collects data for the larger category of smelting and pressing of ferrous metals, it includes steel production enterprises as well as ferroalloy production enterprises.[7]

The sources listed in Table 3 have been the main sources of the statistical data on China energy use for many years. Of the three sources, the two most meaningful are the data from the former Ministry of Metallurgical Industry, currently renamed the China Iron and Steel Association, and the National Bureau of Statistics of China.

Both the *China Steel Industry Yearbook* and the *China Steel Statistics* are managed by the China Iron and Steel Association, and data on national energy consumption are provided by these sources only up to 2003. The *China Energy Statistical Yearbook* is published by the National Bureau of Statistics of China and has more comprehensive data on energy use by the iron and steel industry. However, the *China Energy Statistical Yearbook* provides data on energy use by the larger category of ferrous metal smelting and rolling processing industry which includes iron and steel production enterprises as well as ferrous alloy production.

It should also be noted that Chinese government energy use statistics for the iron and steel industry are based on enterprise information, as stipulated in the corporate law, while energy use statistics gathered by other countries are normally based on the categorization of products, based on product laws. This can be seen in the *China Energy Statistical Yearbook* produced by the National Statistics Bureau of China, in which corporations' enterprise energy use does not correspond to products. This includes not only energy use by the manufacture of iron and steel products but energy use in dealing with other businesses (such as construction, machine repair, transportation, chemical engineering, etc.) conducted by enterprises. In other words, the scope of energy use statistics covered by the *China Energy Statistical Yearbook* is broad and complicated. For example, a company may have been registered as an iron and steel enterprise, but its business lines may cover other industries, such as mining, hotels, construction materials, chemical engineering, transportation, machine repair, schools, restaurants, or hospitals.

In addition, a double-counting occurred in the calculation of the energy use in the Chinese steel industry by National Bureau of Statistics of China. The double-counting involves the accounting for heat and on-site power generation. Section 4.2.2 of this report provides an explanation of the adjustments that were made to the Chinese energy data to account for these issues.

Table 3. Main Data Sources of the Chinese Steel Industry

Main Sources of Energy Data	Main Authorities	Coverage	Contents
China Steel Industry Yearbooks	Managed by Ministry of Metallurgical Industry, National Bureau of Metallurgical Industry and China Iron and Steel Association, successively	Steel industry, key medium-large steel enterprises	• Total energy consumption: national steel industry consumption up to 2003; energy consumption of key steel enterprises for each year, except no data for 2000 and no data collection after 2005 • Energy consumption data by process in key medium-large steel enterprises (source of the data for Table 2) • Production, exports and imports of main products (crude steel, steel products, pig iron, steel billets and steel ingots)
China Steel Statistics (Edited by China Iron and Steel Association, China Steel Statistics, 1991-2009)	Managed by Ministry of Metallurgical Industry, National Bureau of Metallurgical Industry, Metallurgical Information and Standard Research Institute and China Iron and Steel Association, successively	Steel industry	• Total energy consumption data: new version of China Steel Statistics started from 1994, with the same total energy consumption data as China Steel Industry Yearbook. But data only provided through1999. • Production, exports and imports of main products (crude steel, steel products, pig iron, steel billets and steel ingots)
China Energy Statistical Yearbooks	Dept of Industry and Transport, NBS, NEA of NDRC	Smelting and pressing of ferrous metals	• Total energy consumption data: started in 1986, have data collection of energy use in smelting and pressing of ferrous-metals • Imports and exports of steel products

3.1.2. Data Availability – U.S.

The most comprehensive and complete dataset available for iron and steel industry energy use in the U.S is provided in the *Manufacturing Energy Consumption Survey* (MECS) published by the U.S. Department of Energy's Energy Information Administration (EIA) (U.S. DOE/EIA, 2010a-k). The MECS data provide electricity and fossil fuels used in different U.S. manufacturing sectors, including the iron and steel industry. MECS provides energy use of the entire iron and steel industry by fuel type. MECS also disaggregates total energy use by the purpose of the use, i.e. fuel or nonfuel (feedstock).

The MECS data are not disaggregated by the type of steel production technology; hence energy use of EAF steel production and blast furnace - basic oxygen furnace (BF-BOF) steel production cannot be directly obtained from the MECS data. Also, MECS does not provide disaggregated data for energy use in each process step of iron and steel production such as coke making, iron making, steel making, and casting/rolling/finishing. Therefore, it is not possible to calculate the energy intensity for each process step for the whole U.S. iron and steel industry from the MECS data. There are some studies that provide the energy intensity of the U.S. iron and steel industry by process step or by type of technologies used, such as Stubbles (2000). These studies are useful, but the energy intensities provided in them are based on several case studies and not the actual energy use of all the steel plants in U.S.

Production data for the U.S. iron and steel industry are available in the U.S. Geological Survey (USGS) *Minerals Yearbook - Iron and Steel*, which are available for each year. Production data for pig iron, crude steel, finished steel as well as different types of the steel products are available in these reports. American Iron and Steel Institute (AISI) also publishes the *Annual Statistical Report* in which steel production as well as energy use and other data for the U.S. steel industry are given.

Recent information on the overall penetration of different technologies, including energy-efficient technologies, in the U.S. iron and steel industry is not available. MECS does provide information on the number of iron and steel plants that have waste heat recovery, computer control systems for process or major energy-using equipment, and adjustable speed motors. The most recent detailed process-level study of the U.S. iron and steel industry that is publicly available is Stubbles (2000). After this study there are only a few general or industry-level studies that are published, such as AISI (2005).

Recent time series data on the U.S. steel industry are published by the U.S. Department of Energy, the U.S. Geological Survey (USGS), and the

American Iron and Steel Institute (AISI). The AISI publishes periodic industry analyses with useful data such as the *Saving One Barrel of Oil per Ton* roadmap (AISI, 2005).

3.1.3. Base Year for Analysis

Based on the available data, this study focuses comparing energy use and energy intensity of the iron and steel industry in China and the U.S. in 2006.

3.2. Boundaries

In order to make an accurate and fair comparison of the energy intensity of the iron and steel industries in China and the U.S., analytical boundaries must be clearly and accurately defined based on available data in both countries.

There are a number of issues related to defining the boundaries used in this study, including whether to include ferroalloy smelting within the iron and steel industry analysis and how to account for imported and exported products.

3.2.1. Smelting and Pressing of Ferrous Metals

In Chinese statistics, energy use for the smelting and pressing of ferrous metals is included in the overall steel industry energy statistics (category 32 in the NBS industrial categorization (NBS, 2010b). This category is broken down into iron-making (3210), steel-making (3220), steel pressing (3230), and ferroalloy smelting (3240). In the U.S., the *Manufacturing Energy Consumption Survey* (MECS) reports the energy use of iron and steel mills under NAICS category 331111, while the energy use for electrometallurgical ferroalloy products is reported under NAICS category 331112, and the energy use of steel products from purchased steel is reported under NAICS category 3312. Thus, the main difference between the steel industry energy consumption statistics of the two countries is that China's system includes ferroalloy smelting while energy use for this activity is provided separately in the U.S. statistics (see Table 4).

Currently the international community only recognizes the steel industry as steel production enterprises. As such, for China this report will be based on energy consumption data in the category of "smelting and pressing of ferrous metals" as published in China Energy Statistical Yearbook in 2006, with the energy consumption of ferroalloy smelting deducted from the total. For the U.S., this report will use the energy consumption of iron and steel mills (NAICS 331111) and steel products from purchased steel (NAICS 3312) since they are within the boundary of this study. The energy use of electrometallurgical ferroalloy products (NAICS 331112) is not included in the calculation of the energy intensity in this study.

Table 4. Comparison of the Definitions and Scope of the Iron and Steel Industry in China

Country	Steel Industry		Scope		
China	Smelting and pressing of ferrous metals (Industry code 32)	Iron-making (Industry code 3210)	Steel-making (Industry code 3220)	Steel pressing (Industry code 3230)	
US	Iron and steel mills (NAICS 331111)	Iron-making (DRI, blast furnace)	Steel-making	Steel rolling	Steel products frompurchased steel (NAICS 3312)

3.2.2. Accounting for Imported and Exported Products

In China, about two thirds of consumed coke in the steel industry is produced separately by independent coking plants, and another third of the coke is produced by steel mills themselves. Net exports of main steel products (pig iron, steel billets, and steel products) are about 8% of total crude steel production (after converting them to crude steel production).

In the U.S., the net imported (import minus export) pig iron in 2006 was 5.92 Mt, net imported Direct-Reduced Iron (DRI) in 2006 was 2.61 Mt, and net imported ingots, blooms, billets and slabs in 2006 was 8.26 Mt. The total net imported steel mill products were about 32.27 Mt in 2006 (USGS 2008).

When one enterprise or one country does not have enough upstream production capacity, it needs to purchase upstream products such as pig iron, coke, and DRI. Energy consumption for production of these products is also counted in the total energy consumption in this study. On the other hand, when there is a surplus in upstream production capacity and a portion of the products are sold, the energy consumption of these sold products is deducted from total energy consumption in this study.

Besides the main processes, Chinese steel enterprises also include auxiliary systems, such as oxygen preparation, while most U.S. steel companies do not include auxiliary systems but instead purchase oxygen and other gases. However, the energy use for the production of purchased oxygen is taken into account in the calculation of energy intensity in both countries.

3.2.3. Boundaries Used in This Study

In this study, the boundaries of the iron and steel industry include all coke making, pelletizing, sintering, iron making, steel making, steel casting, hot rolling, cold rolling, and processing such as galvanizing or coating (Figure 3). This boundary definition is used for the calculations of energy use and energy intensity for both the Chinese and U.S iron and steel industries. This study takes net imported (to the boundary of the industry defined in Figure 3) pig iron, direct-reduced iron (DRI), pellets, lime, oxygen, and ingots, blooms, billets, and slabs into account by adding the energy used for production of these products to the total energy input to the iron and steel industry.

Regarding accounting for energy used for coke production within the iron and steel industry, there are a few special considerations. This study includes the total coal input used as a feedstock for coke making and also as a fuel in other parts of the steel making process. Only net imported coke (either produced in other domestic industries or imported from other countries) is included as a source of input energy to the iron and steel industry. Net imported coke is total imported coke minus total exported coke. The energy value of the coke produced in the coke making process within the iron and steel industry and used in the iron making process is not included since the coal initially used to produce the coke is already accounted for within the boundary. This study does not count the coke trade that occurs within the boundary, as the total coal input to the industry is already taken into account.

This study takes net imported (to the boundary of the industry defined in Figure 3) pig iron, direct-reduced iron (DRI), pellets, lime, oxygen, and ingots, blooms, billets, and slabs into account by adding the energy used for production of these products to the total energy input to the iron and steel industry.

In addition, this study does not include energy consumption associated with other energy-intensive products manufactured for the industry (e.g., electrodes, ferroalloys, refractories, etc.). These products could be included in a more extensive, life-cycle analysis study of the industry, but are excluded here because the focus of this study is on iron and steel production. This is the approach taken by Stubbles (2000). This study also does not take into account the embodied energy of the scrap used in the iron and steel industry. In other words, the embodied energy of the scrap is not added to the energy use of the EAF process. Also, energy demand for mining and beneficiation of iron ores is not included in this analysis. Finally, the energy use in further processing of the steel by steel foundries is not included in this analysis.

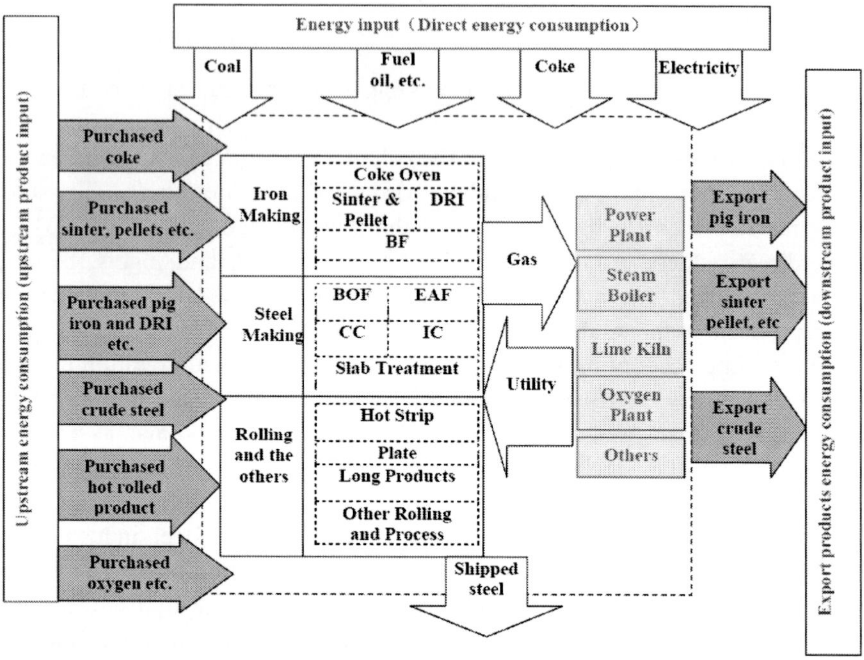

Figure 3. Flowchart of Iron and Steel Sector Boundaries Used in this Study.

3.3. Conversion Factors

In order to calculate comparable energy use and energy intensity values, common conversion factors must be used to convert the physical quantities of fuels consumed to produce steel to energy values. In addition, common conversion factors must also be used to calculate electricity values. These conversion factors are explained below.

3.3.1. Fuel Conversion Factors

The heating value or calorific value of a fuel source represents the amount of heat released during combustion. This study uses the lower heating value – or net calorific value (NCV) – to convert physical quantities of fuel to a common energy unit. NCV conversion factors for China are provided in the *China Energy Statistics Yearbook* (2007, 2009) and for the U.S. are provided in the U.S. Energy Information Administration's *Annual Energy Review* (2007, 2009). Where available, the NCV of the fuels in 2006 is used. In addition, the typical NCVs for fuels are also provided in various IEA publications. Table 5 provides the NCV conversion factors for different fuels for China and the U.S. Figure 4 shows a graphical comparison of the NCVs from the different sources.

From Table 5 it can be seen that the fuel conversion factors in the two countries are different, especially those for other bituminous coal and coking coal. Small differences in conversion factors might cause great differences in total energy consumption. For example, China's steel industry consumed 104.07 Mt of washed/cleaned coal (wet-base, with ~10% moisture). Based on the Chinese conversion factor of 1.0143 kgce/kg (29.727 MJ/kg), cleaned coal (dry-base), energy consumption is 95 Mtce (3093.66 PJ). If converted using the U.S. conversion factor of 1.039 kgce/kg (30.451 MJ/kg) cleaned coal, energy consumption is 97.32 Mtce (3168.998 PJ). The difference between the calculated energy consumption is 2.32 Mtce (75.338 PJ).

Table 5 shows that:

1. The IEA recognizes that energy quality in different countries varies, and to reconcile the difference and compare internationally, IEA proposes a unified system of energy conversion factors. These average values are applicable to all countries. Since one of the goals of this study is to develop a methodology that can be used to compare the iron and steel industries of different countries in the world, one scenario is developed in which the IEA typical fuel NCVs are used.
2. The main difference between the energy conversion factors in the U.S. and China is for other bituminous coal and clean coal. The NCV used in the U.S. is for dry-base coal, but the NCV used in China is for wet-base coal. For this analysis, however, the dry-base conversion factor is used for China.

For this study, the international average energy conversion factors are used for products that are purchased externally and imported or exported by the iron and steel industry since imported products can be from different countries and will thus vary in their energy consumption during production due to differences in production technology and energy structure. The energy conversion factors for external products in this study are provided by the World Steel Association (worldsteel) (worldsteel, n.d.; worldsteel, 2008b).

Table 5. Fuel Conversion Factors for China and the U.S.

Fuel	IEA-Typical		Country-Specific				Unit
	IEA-Typical	Source	China	Source	U.S.	Source	
Other Bituminous coal (used as fuel)	24.05 [a]	IEA 2005	20.91	NBS 2007	25.65	EIA 2009	MJ/kg
Coking coal	28.20 [b]	IEA 2005	26.34	NBS 2007	30.56	EIA 2009	MJ/kg
Coke oven coke	27.45	IEA 2005	28.44	NBS 2007	28.85	EIA 2009	MJ/kg
Natural gas	35.04 [c]	IEA 2008c	38.93	NBS 2007	38.33	EIA 2009	MJ/m3
Residual Fuel oil	42.18 [d]	IEA 2005	---		44.18	EIA 2007	MJ/kg
Distillate Fuel Oil	40.19	IEA 2008c	41.82	NBS 2007	40.94	EIA 2007	MJ/kg
LPG	46.15 [e]	IEA 2005	50.18	NBS 2007	45.81	EIA 2009	MJ/kg
Other washed coal	-		10.47 [g]	NBS 2007			MJ/kg
Crude oil	42.85 [f]	IEA 2008c	41.82	NBS 2007			MJ/kg
Gasoline	47.10	IEA 2005	43.07	NBS 2007			MJ/kg
Kerosene	46.22	IEA 2005	43.07	NBS 2007			MJ/kg
Diesel	45.66	IEA 2005	42.65	NBS 2007			MJ/kg
Other petroleum products	-		35.17	NBS 2007			MJ/kg
Tar	-		33.45	NBS 2009			MJ/kg
Benzene			41.82	NBS 2009			MJ/kg

[a] IEA provides a range for typical NCVs of other bituminous coal of 22.6 to 25.5 MJ/kg. The average value of 24.05 MJ/kg (0.821 kgce/kg) is used in the table above.

[b] IEA provides a range for typical NCVs of coking coal of 26.6 to 29.8Mj/kg. The average value of 28.2 MJ/kg (0.962 kgce/kg) is used in the table above.

[c] Natural gas as supplied contains gases in addition to methane (usually ethane and propane). As the heavier gases raise the calorific value per cubic meter, the gross calorific values can vary quite widely. Therefore, the average NCV of natural gas of the top ten largest producers in 2008 is used in the table above.

[d] This is for low sulfur fuel oil.

[e] IEA value for LPG is a typical value and is the same for both countries.

[f] The average NCV of crude oil of the top ten largest producers in 2008 is used in the table above.

[g] Average value of 8.37 and 12.56 MJ/kg.

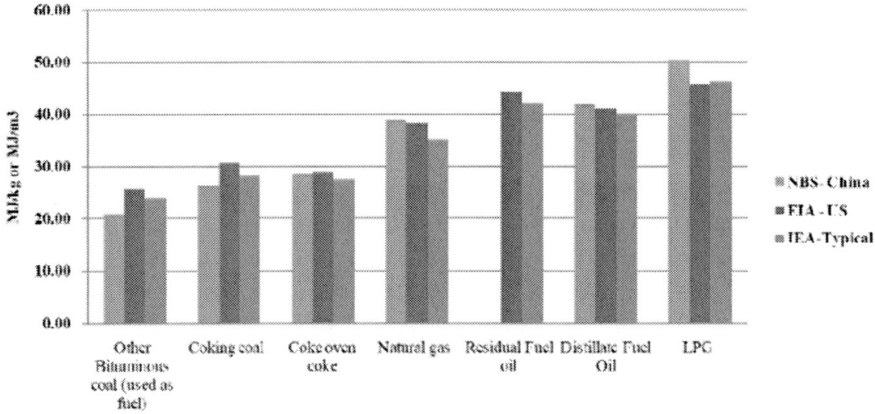

Figure 4. Fuel Conversion Factors for China and the U.S.

Table 6. Conversion Factors for Purchased Fuels & Auxiliary/Intermediary Products

	Coke	Pig Iron[a]	Coal based[a] DRI	Gas based [a] DRI	Pellets [a]	Crude Steel[b]	Lime	Oxygen[a]
	MJ/kg	MJ/kg	MJ/kg	MJ/kg	MJ/kg	MJ/kg	MJ/kg	MJ/m^3
worldsteel Factors (Primary Energy)	4.0	20.9	17.9	14.1	2.1	18.9	4.5	6.9
China-specific value (Primary Energy)						18.54		
worldsteel Factors (Final Energy)[c]	3.7	19.8	17.0	13.4	2.1	16.5	4.1	2.5
China-specific value (Final Energy)						17.4		
Electricity share in total Primary Energy	11%	8%	8%	8%	0	20%	15%	100%
China-specific value for electricity share in total Primary Energy					0	10%		

[a] worldsteel, n.d.; [b] worldsteel, 2008a; [c] The 9.8 MJ/kWh conversion from worldsteel was used to convert the worldsteel conversion factors for Purchased Fuels & Auxiliary/Intermediary Products from primary to final energy using the percentages of electricity use for the production of each product given in the table above.

The energy conversion factors for all products except for crude steel provided by the worldsteel.org are in primary energy[8]. Hence, for all products except for crude steel, the energy conversion factors are converted into final energy for the final energy intensity calculation and only the crude steel energy conversion factor is converted from final to primary energy (see Table 6). For this purpose, the share of electricity use in the production of auxiliary and intermediary products and the conversion factor for electricity to primary energy is used.

Table 6 provides energy conversion factors for purchased fuels and materials as well as imported auxiliary/intermediary products along with the share of electricity use for production of each product. The values provided by worldsteel are assumed to be the international average; thus these values are used for the base case of this study. However, there are also U.S. country-specific values mostly from Stubbles (2000).

3.3.2. Electricity Conversion Factors

Final (or site) electricity is the electricity consumed at the production facility. This value, however, does not include the primary energy used to generate electricity (both offsite and onsite) and transmit and distribute (only by offsite power producers) electricity to the site. To convert final electricity to primary energy, the average efficiency of power generation and transmission and distribution (T&D) losses must be taken into account. Primary energy is used in the energy conversion system (power sector) to produce electricity. Since the electricity generation has a certain efficiency which is often in the range of 25% - 40%, one unit of electricity in final energy terms is always less than the actual primary energy used to produce that unit of electricity. Total primary energy use is the sum of fuel use and the electricity use that is converted to primary energy by taking into account the power generation efficiency and T&D losses.

Presenting energy intensity in primary energy terms is especially important for the iron and steel industry because the EAF process uses mostly electricity, whereas the BF-BOF process uses mostly fossil fuels. Hence, if energy intensity is just compared in terms of final energy, it underestimates the actual primary energy needs of the industry since it ignores the large energy losses during electricity generation and the losses associated with T&D of electricity from offsite sources. EAF steel accounted for 56.9% of the steel produced in the U.S in 2006, whereas it accounted for just 10.5% of the steel produced in China in the same year (worldsteel, 2009). Therefore, comparing the energy intensities in primary energy is important since there is such a difference in the share of EAFs in China and the U.S.

In this study, the average heat rate of fossil fuel-fired power generation is used to convert electricity from final to primary energy since fossil fuels are the dominant sources of energy used for power generation in both China and the U.S. Heat rates can be expressed as either gross or net heat rates, depending whether the electricity output is gross or net generation. Net electricity output is the electricity after the deduction of power plant self-use. Heat rates are typically expressed as net heat rates; hence in this study the net heat rate is used. The average net heat rate of fossil fuel-fired power generation in China and U.S in 2006 is given in Table 7. A significant amount of electricity is also lost in the transmission and distribution (T&D) grid, which should be included when converting electricity from final to primary energy. Table 8 provides the grid T&D losses in China and the U.S. in 2006.

Table 7. Average Net Heat Rates of Fossil Fuel-Fired Power Plants in 2006

	Net Generation Heat Rate	Unit	Source
China	0.350	kgce/kWh	NBS 2009
U.S	0.354	kgce/kWh	EIA 2010a
worldsteel	0.334	kgce/kWh	worldsteel2008b

Note: Coal-fired power generation in China is calculated based on thermal power generation standard coal and thermal capacity provided in the *China Energy Statistical Yearbook* 2009.

Table 8. Grid Transmission & Distribution Losses in China and the U.S. in 2006[9]

Country	Grid Transmission & Distribution Losses	Source
China	7.0%	Anhua and Xingshu, 2006
U.S	6.7%	EIA 2008

Using the net generation heat rate and T&D losses from the tables above, electricity can be converted from final to primary energy using the following formula:

$$CF = NHR / (1 - T\&D \text{ losses})$$

Where:

CF: Conversion factor to convert electricity from final to primary energy (kgce/kWh)

NHR: Average net heat rates of fossil fuel-fired power plants (kgce/kWh)

T&D losses: Grid T&D losses (%)

The conversion factors to convert electricity from final to primary energy for the U.S. and China in 2006 that are calculated using the above tables and formula are presented in Table 9. By multiplying the final electricity by this conversion factor, the electricity in primary energy terms can be obtained. As can be seen from Table 9, the electricity conversion factors for the two countries are close to each other.

Table 9. Final to Primary Energy Conversion Factor in 2006

	China	U.S
Final Conversion FactorwithT&D losses(kgce/kWh)	0.376	0.379
Final Conversion Factor withoutT&D losses (kgce/kWh)	0.350	0.354

4. BASE YEAR PRODUCTION, TRADE AND ENERGY USE DATA

4.1. Production and Trade Data

4.1.1. Production Data for the U.S.

Table 10 shows the production and trade data for pig iron, DRI, crude steel, ingots, blooms, billets, slabs, and steel products (finished steel). As can be seen from this table, net import of pig iron accounts for 13.5% of the total pig iron used in the U.S. This is a substantial amount which cannot be neglected. For the calculation of the energy intensities, crude steel production is used as the denominator. However, it should be noted that the casting, rolling and finishing processes that happen after the crude steel production are also within the boundary of the analysis.

Table 10. Production and Trade Data for Pig Iron, DRI, Crude Steel, Ingot, Blooms, Billets, and Slabs, and Steel Products in U.S in 2006 (Mt)

Product	Production	Exports	Imports	Net Imports	Used in industry
Pig Iron	37.9	0.813	6.73	5.92	43.8
DRI	0.24	-	2.61	2.61	2.85
Crude Steel	98.2	-	-	-	-
Ingots, Blooms, Billets, Slabs	-	0.20	8.46	8.26	-
Steel Products	99.3	8.83	41.1	32.3	-

Source: USGS, 2008.

4.1.2. Production and Trade Data for China

Table 11 lists China's production, exports and imports of pig iron, DRI, crude steel, and steel products in 2006. Net exports are 0.7 Mt of pig iron and 8.67 Mt of steel billets. Net imports are 0.3 Mt of DRI, and 0.1 Mt of steel ingots.

Table 11. Production, Imports and Exports of Pig Iron, DRI, Crude Steel, Ingots, Billets, and Steel Products in China, 2006 (Mt)

Product	Production	Exports	Imports	Net Trade	Used in industry
Pig Iron	413.64	0.87	0.17	-0.70	412.94
DRI	0.21	0.01	0.31	0.30	0.51
Crude Steel	421.02	-	-	-	-
Steel Ingots	-	0.04	0.14	0.10	-
Steel Billets	-	9.04	0.37	-8.67	-
Steel Products	399.97 [a]	43.01	18.51	-24.50	375.47

Source: Editorial Board of the China Iron and Steel Industry Association Yearbook, 2006.

* In order to avoid double-counting of steel products, this number was calculated as 95% of crude steel.

4.2. Energy Use Data

4.2.1. Energy Use of the U.S. Iron and Steel Industry Based on EIA Reported Data

U.S. energy use data in *physical units* are presented in Table 12. This table provides energy use for fuel and nonfuel (e.g. a feedstock such as coke), defined as "first use" of energy by the EIA (U.S. DOE/EIA, 2010b). Table 13 provides first use of energy for all purposes (fuel and nonfuel) in *energy units* for the U.S. iron and steel industry in 2006 based on the EIA fuel conversion factors from U.S. DOE/EIA (2010e, f, g). It should be noted that the data in Tables 12 and 13 are in final energy. Therefore, the electricity data given in these two tables do not include the primary energy used to generate electricity (both offsite and onsite) and transmit and distribute (only by offsite power producers) electricity to the site.

In addition to the first use of energy, which is the energy used as fuel and nonfuel (feedstock), the energy use for the production of net imported coke, lime, pellets, pig iron, DRI, oxygen, and ingots, blooms, billets, and slabs will also be included in the calculation of energy intensity for this study in order to have a more accurate and fair comparison of the energy intensity of the industry in both countries. This is done to eliminate the effect of differences in the share of imported coke, lime, DRI, and pig iron on the energy intensity of the industry in the two countries. For the base case scenario of this study, the worldsteel conversion factors for these auxiliary/intermediary products consumed in the iron and steel industry is used. The 9.8 MJ/kWh conversion factor from worldsteel (n.d.) is used to convert the worldsteel conversion factors for purchased fuels and auxiliary/intermediary products from primary to final energy using the percentages of electricity use for the production of each product given in Table 6.

Worldsteel (n.d.) provides an energy intensity of 3.7 GJ/tonne for the production of coke, 4.1 GJ/tonne for lime, and 19.8 GJ/tonne for pig iron. These values were assumed to be world average energy intensities for the production of coke, lime, and pig iron and are used in the base case scenario of this study. By multiplying the net imported coke (3,108,669 tonnes) and lime (1,956,000 tonnes) from both outside of the U.S. as well as from other industries in the U.S. into the steel industry by the energy intensity of coke and lime production, the amount of energy that is associated with the production of net imported coke and lime are calculated. These values are added to the energy use before the final energy intensity calculation. It is assumed that 89% of total energy use for coke production is fuel, and 85% of total energy use for lime production is fuel. The amount of net imported lime (i.e. lime produced outside the iron and steel industry and used in the iron and steel industry) for the U.S. is obtained from USGS (2007).

Pelletizing is the primary ore agglomeration technology, but it normally takes place at the mine site. It is therefore not considered as part of the iron and steel industry according to the NAICS classification (U.S. Census Bureau, 2007). Therefore, the energy use given for the iron and steel industry in the

U.S. does not include pelletizing, but Chinese energy statistics includes pelletizing in iron and steel energy use data. To include the amount of energy used for the production of pellets in the U.S., the following calculation is made. The amount of pellets used in the U.S. steel industry in 2006 was 49.30 Mt (USGS, 2008). The energy use for the pelletizing is 2.1 GJ/tonne (worldsteel, n.d.). Thus, energy use for the pellets used in the U.S. iron and steel industry is calculated and added to the fuel use. It is assumed that all the energy use for the pellet production is fuel, so this value is added to the total fuel consumption. Based on NAICS, sintering is included in the iron and steel industry and, assuming there is no sinter imported from outside of the U.S., energy use for the sintering process is thus included in the energy use given by MECS. Sintering accounted for only 12% of iron oxide used in the U.S. iron and steel industry in 2006. The rest of the iron oxide used was pellets (USGS, 2008).

Furthermore, by multiplying the net imported pig iron and DRI by the energy intensity of pig iron and DRI production, the amount of energy that is associated with the net imported pig iron and DRI can be obtained, respectively. This value should be added to the first use energy use before the final energy intensity calculation. It is also assumed that around 92% of the energy use for pig iron as well as DRI production is fossil fuels and the remaining 8% is the electricity use.[10] There are two types of processes for DRI production: coal-based and natural gas-based. Based on Chukwuleke et al. (2009) it is assumed that 85% of the net imported DRI is produced by the natural gas-based process and 15% by the coal-based process. The calculated electricity and fossil fuel use for the production of net imported pig iron and DRI was added to the first use of electricity and fossil fuel in the U.S iron and steel industry given in Table 12. Also, the boundary of this study includes the energy consumption for the production of the net imported ingots, blooms, billets, and slabs as well as the energy use for rolling and finishing processes of net imported ingots, blooms, billets, and slabs. In 2006 the U.S. imported a net amount of 8.261 Mt ingots, blooms, billets, and slabs. These net imported ingots, blooms, billets, and slabs are taken into account in the analysis as described below.

In MECS 2006, the NAICS category 3312 represents the energy use for the steel products from purchased steel (i.e. imported ingots, blooms, billets, and slabs). This is the energy used for the rolling and finishing processes conducted on the imported ingots, blooms, billets, and slabs. The net imported ingots, blooms, billets, and slabs in 2006, were 8.261 Mt. Based on worldsteel (2008a, b) the final average energy intensity of steel production is 20.6

GJ/tonne crude steel. This includes the entire finished steel production process, i.e. crude steel production as well as rolling and finishing processes. Since only the energy intensity of crude steel production is needed, the portion of the energy intensity related to crude steel production (excluding rolling and finishing processes) can be calculated from the value given above (20.6 GJ/tonne crude steel). Based on Stubbles (2000), around 80% of the total energy use for the production of finished steel is for crude steel production in both EAF and BF-BOF production lines. By multiplying the 20.6 GJ/tonne crude steel by 80%, the energy use for the production of crude steel is calculated to be 16.5 GJ/tonne crude steel. In addition, having the EAF ratio of the countries from worldsteel (2009) and the amount of ingots, blooms, billets, and slabs imported from each country in 2006 from USGS (2008) the weighted average ratio of EAF production in the countries from which the U.S. imported ingots, blooms, billets, and slabs in 2006 was calculated and is equal to 37.1%. Based on data provided in Stubbles (2000) it is assumed that 95% of the total energy use in BF-BOF plants is fuel, while in EAF plants around 55% of the total final energy use is fuel. Having the share of fuel used in BF-BOF and EAF production lines and the ratio of each, the combined share of fuel for the imported ingots, blooms, billets, and slabs was calculated to be around 80% and the other 20% is the electricity use. These shares are used for the calculating the embodied energy of the imported ingots, blooms, billets, and slabs into electricity and fuel.

In 2006, the U.S. steel industry purchased 6,786 million m^3 of oxygen and produced 194 million m^3, for a total oxygen use of 6,980 million m^3 (AISI, 2008). Table 14 below shows the production, consumption and purchased oxygen in the U.S. iron and steel industry by process. worldsteel (n.d.) gives an energy intensity of 0.7 kWh/m^3 of oxygen. Having the amount of purchased oxygen and the energy intensity of the oxygen production, the energy consumption for the production of purchased oxygen was calculated and added to the energy use of the iron and steel industry.

The total electricity and fuel consumption for the production of iron and steel in U.S. based on the defined boundary of this study is presented in Table 15.

Table 12. Energy Use (Fuel and Nonfuel) in Physical and Energy Units for the U.S. Iron and Steel Industry, 2006

NAICS	Industry	Electricity[1]	Residual Fuel Oil	Distillate Fuel Oil	Natural Gas	Coal used as fuel (calculated from EIA, 2010b)[2]	Coal used for coking (calculated from EIA, 2010b)[2]	Coke and Breeze	Other[3]	Shipments of Energy Sources Produced Onsite[4]	Total[5]
		million kWh	million liters	million liters	billion m^3	million tonnes	million tonnes	million tonnes	TJ	TJ	TJ
331111	Iron and Steel Mills	51,198	477	159	10	0.7	14.1	8.2	41,147	152,983	1,096,936
3312	Steel Products from Purchased Steel	4396	7.5	0	0.7	0.02	0	0.01	14,771	0	59,083

Source: U.S. DOE/EIA, 2010b. Note: Energy data are for the North American Industry Classification System (NAICS) categories of 331111 and 3312.

[1] Electricity use data are in final energy. That is, the electricity use is not converted to the primary energy use by taking into account the power generation efficiency and transmission and distribution losses.

[2] Total first use of coal is 16 million short tons in U.S. DOE/EIA (2010b). "Coal use for fuel" in U.S. DOE/EIA (2010c) is 1 million short tons and coal use as nonfuel (for coking) in U.S. DOE/EIA (2010d) is 16 million short tons, giving a total of 17 million short tons. This value is higher than the total value given in U.S. DOE/EIA (2010b) due to rounding. However, since 1 million short tons is a significant amount of coal, this difference is not acceptable for the analysis conducted in this study. To get the same rounded values as those given in U.S. DOE/EIA (2010b, c, d), total coal use was recalculated using the energy values (U.S. DOE/EIA, 2010 e, f) and the NCV for each type of coal used (coal used as fuel and coal used for coking[11]). The values calculated in this manner for coal used as fuel were 0.8 million short tons (0.7 million metric tonnes) and for coal used as nonfuel (for coking) were 15.6 million short tons (14.1 million metric tonnes), for a total value of 16.4 million short tons, or 14.8 million metric tonnes.

[3] The unit used for the category “Other” is an energy unit instead of a physical unit since this category is comprised of several fossil fuels which are not specified in U.S. DOE/EIA (2010b). Also, for the category “Shipments of Energy Sources Produced Onsite”, EIA (2010b) gives the values in energy units.

[4] The category “Shipments of Energy Sources Produced Onsite” is for those shipments produced or transformed onsite from the nonfuel use of other energy sources. For example, at an establishment that processes coal to make coke for later use, the entire quantity of coal is counted as first use. Any onsite consumption of coke is not counted as first use because it would duplicate the coal use. If some of the coke is then sold to another establishment, then that second establishment will consider this coke to be a shipment of an offsite-produced energy source. Hence, the second establishment will count this coke as its first use, thereby resulting in double counting. In order to eliminate double counting, the energy equivalent of the coke shipment must be subtracted from first use. In other words, the total first use energy consumption of the U.S. iron and steel industry in Table 12 and Table 13 can be obtained by summing up the energy value of the electricity and all the fossil fuel used and then subtracting the “Shipments of Energy Sources Produced Onsite”.

[5] Total energy consumption is calculated by EIA using the energy used in physical units and multiplying it by the net calorific value (NCV) of each fossil fuel. For electricity, only the unit conversion factor is used to convert electricity from million kWh to terajoules (TJ).

Table 13. Energy Use for All Purposes (Fuel and Nonfuel) for the U.S. Iron and Steel Industry Using EIA Fuel Conversion Factors, 2006 (TJ)

NAICS	Industry	Electricity	Residual Fuel Oil	Distillate Fuel Oil	Natural Gas	Coal used as Fuel[1]	Coal used for Coking[1]	Coke and Breeze	Other	Shipments of Energy Sources Produced Onsite	Total
331111	Iron and Steel Mills	184,313	20,046	3,165	378,765	14,771	365,049	242,663	41,147	152,983	1,096,936
3312	Steel Products from Purchased Steel	15,826	317	0	27,431	422	0.0	317	14,771	0	59,083

[1] Calculated from EIA, 2010b.

Table 14. Production, Consumption and Purchases of Oxygen in the U.S. Iron and Steel Industry in 2006

	2006 (Million m^3 in gaseous form)
Purchased	6,786
Produced	194
Total	6,980
Used in:	
Blast Furnace	2,329
Steelmaking:	
BOF	2,458
EAF	1,966
Total BF-BOF	4,786
Total EAF	1,966
All other	227
Total iron and steel industry	**6,980**

Source: AISI (2008).

4.2.2. Energy Use Data for the Chinese Iron and Steel Industry

Table 16 provides the end-use final energy consumption in physical units of the category of Smelting and Pressing of Ferrous Metals which also includes the energy use of ferroalloys smelting in China in 2006. The energy consumption shown in this table also includes double-counted secondary energy for heating. In order to derive the energy use of the primary fuels used by China's iron and steel industry as reported in Table 17 below, the following adjustments to the data must be made:

- *Remove On-Site Generated Electricity.* The electricity use shown in Table 16 is the total electricity use in the industry, but not all of it is the purchased electricity. Iron and steel corporations tend to install electric voltage themselves and are equipped with energy efficiency technologies such as CDQ and TRT, as well as generate electricity on their own up to a certain percentage.[12] To avoid double counting, the onsite generation of electricity is taken out of the total electricity use and the resulting purchased electricity amount is shown in Table 17.
- *Convert Non-Purchased Coke to Cleaned Coal.* In Table 16 the total coke used in the Chinese Smelting & Pressing of Ferrous Metals (including ferroalloys) in 2006 is given. Part of this coke is produced within the iron and steel industry boundary defined in this study and the rest is purchased from outside of the industry. The portion that is purchased from outside of the industry boundary (137.19 Mt) is given

in Table 17 and the portion of coke that is produced within the boundary (98.67 Mt) is converted back to cleaned coal (coking coal) and added to the rest of the clean coal used in the industry and presented in Table 17.

Table 15. Total Electricity and Fuel Consumption for Iron and Steel Production in the U.S. Based on the Study Bounda

Component	Electricity Use (GWh)	Fuel Use (TJ)	Final (TJ)	Primary (TJ) *
Energy use reported for the iron and steel industry in EIA *(excluding the energy use for production of intermediary products given below)*	51,198	912,623	1,096,936	1,481,942
Energy used for the production of net imported oxygen	4,750	0	17,101	52,824
Energy used for the production of net imported pig iron	2,603	107,784	117,157	136,735
Energy used for the production of net imported direct reduced iron	809	33,473	36,383	42,463
Energy used for the rolling and finishing of net imported ingots, blooms, billets, and slabs	4,396	43,257	59,083	92,141
Embodied energy of net imported ingots, blooms, billets, and slabs	7,509	109,109	136,141	192,608
Energy used for the production of net imported coke	351	10,237	11,502	14,145
Energy used for the production of net imported lime	334	6,816	8,019	10,532
Energy used for the production of net imported pellets	0	103,530	103,530	103,530
Total energy consumption of steel industry with embodied energy of net imported/exported auxiliary/intermediary products included	71,951	1,326,830	1,585,853	2,126,919

* In final energy, electricity use is equal to the electricity consumption at the end-use. In primary energy with T&D losses, electricity use at the end-use is converted to the primary energy sources by taking into account the power generation efficiency (average net heat rate of power plants) and transmission and distribution losses.

- *Convert Coke Oven Gas (COG) and Other Coal Gas back to Cleaned Coal.* Table 16 provides information on the end use of the by-products of cleaned coal produced by the iron and steel industry, such as coke oven gas and other coal gas, which are normally produced by the iron and steel enterprises and then are used internally, rather than being purchased from external vendors. The China Energy Statistical Yearbook presents only the physical quantity of energy use by end-users (steel plants) for these by-products, not the cleaned coal initially consumed for their production. Thus, these by-products are converted back to cleaned coal and added to the reported clean coal used as a fuel in the calculation of the iron and steel industry's energy use presented in Table 17. Thus, the values of 31.38 billion m^3 for COG, 79.14 billion m^3 for other coal gas, and 1.48 Mt for other coking products are converted to cleaned coal and added to the reported end-use value of 6.82 Mt of cleaned coal and the calculated value for clean coal used to produce coke (explained above) for a total of 135.79 Mt of cleaned coal in Table 17.

Table 16. End-Use Final Energy Consumption (Physical Amount) of Smelting & Pressing of Ferrous Metals (including Ferroalloys) in 2006

Energy type	Unit	Ferrous metal smelting and rolling processing industry
Raw coal	Mt	58.18
Cleaned coal	Mt	6.82
Other washed coal	Mt	1.58
Coke	Mt	235.86
Crude oil	Mt	0.001
Gasoline	Mt	0.25
Kerosene	Mt	0.02
Diesel	Mt	0.92
Fuel oil	Mt	1.51
LPG	Mt	0.13
Other petroleum products	Mt	0.09
Natural gas	109 m^3	1.12
Electricity	10^9 kWh	303.90
Coke oven gas (COG)	109 m^3	31.38
Other coal gas	109 m^3	79.14
Other coking products	Mt	1.48
Heating	10^9 kJ	162,209.1

Source: NBS, 2010a. Note: Mt = million tonnes.

- *Remove Double-Counted Heating Energy.* The amount of heating energy reported separately in Table 16 is the by-product of other processes for which the coal or coke used is already accounted for. Therefore, the heating reported in Table 16 is deducted from the total energy use reported in Table 17.
- *Remove Energy for Ferrous Alloy Production.* Since the smelting and pressing of ferrous metals industry in China includes enterprises of

iron and steel production and ferroalloy production, the energy use by the ferroalloy enterprises is deducted from the total energy use given in Table 16. This energy use is based on statistics provided to the China Iron and Steel Research Institute (CISRI) by China's National Bureau of Statistics (NBS) which show the energy use by China's ferrous alloy industry equal to 1,142,704 TJ. This energy use is deducted from the total energy use shown in Table 16.This value has been confirmed by Chinese experts from the ferrous alloy industry and been deemed reasonable.[13]

After the adjustments described above are made, the revised physical quantity of energy use by the iron and steel industry in China is shown in Table 17.

Table 17. Physical Quantity and Related Energy of Energy Use by the Iron and Steel Industry in China (2006)

Fuel Type	Unit	Physical Quantity	Energy (TJ)
Raw Coal	Mt	53.43	1,117,198.17
Cleaned Coal	Mt	135.79	3,576,761.16
Other Cleaned Coal	Mt	1.27	13,245.42
Purchased Coke	Mt	137.19	3,901,671.83
Crude Oil	Mt	0.0013	54.53
Gasoline	Mt	0.24	10,452.86
Kerosene	Mt	0.02	723.58
Diesel	Mt	0.89	37,879.13
Fuel oil	Mt	1.49	62,477.21
LPG	Mt	12.73	6,385.79
Other petroleum products	Mt	0.03	1,179.04
Natural gas	109 m^3	1.08	42,011.60
Tar Oil	Mt	4.00	-42,659.89
Benzene	Mt	1.02	-133,822.36
Purchased electricity (final)	TWh	174.29	627,456.39
Total (final)			9,221,014.46

Source: China Iron and Steel Research Institute (CISRI) calculations.

Notes:

(1) The energy values reflect the deduction of ferroalloy energy use, double-counted heating energy use, and on-site electricity production.

(2) The fuel conversion factors used to calculate this energy values in table are based on Chinese NCVs

(3) Mt = million tones.

According to the boundaries presented in Figure 3, the energy consumption of steel production is calculated and included in this analysis. The upstream energy consumption of net imported coke, pig iron, DRI, steel ingots and steel billets is presented in Table 18.

Total energy use is adjusted for net trade in auxiliary and intermediate products. Figure 5 provides an illustration of these calculations.

Table 18. Total Energy Consumption of China's Steel Industry Production in 2006

Component	Electricity Use (GWh)	Fuel Use (TJ)	Final (TJ)	Primary (TJ) *
Reported energy consumption *(excluding the energy use for production of intermediary products given below)*	174,293	8,593,558	9,221,013	10,515,967
Energy used for the production of purchased coke	5,883	488,395	509,574	553,283
Energy used for the production of net exports of pig iron	-114	-13,412	-13,822	-14,669
Energy used for the production of net imports of coal-based DRI	42	4,934	5,085	5,397
Energy used for the production of net imports of steel ingots	17	1,589	1,650	1,776
Energy used for the production of net exports of steel billets/slabs	-2,082	-192,304	-199,799	-215,268
Total energy consumption of steel industry with embodied energy of net imported/exported auxiliary/intermediary products included	178,039	8,882,760	9,523,701	10,846,487

* In final energy, electricity use is equal to the electricity consumption at the end-use. In primary energy with T&D losses, electricity use at the end-use is converted to the primary energy sources by taking into account the power generation efficiency (average net heat rate of power plants) and transmission and distribution losses.

Note 1: The negative values indicate the energy use by export products was subtracted.

Note 2: The reason that there is no energy use data given separately for lime and pellets is that the energy use for the production of these products is included in the reported energy consumption of the steel industry in China (first row of this table) and there is no import or export of these two products.

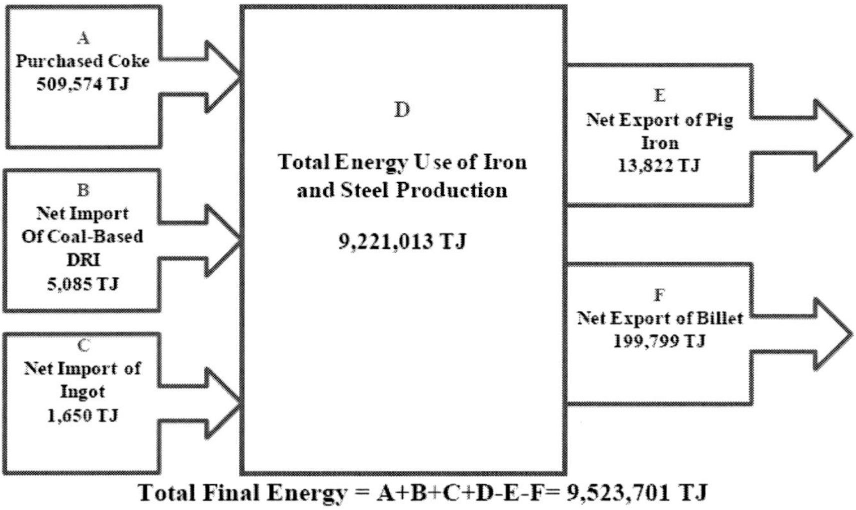

Figure 5. Final Energy Use of China's Iron and Steel Industry in 2006.

5. Comparison of Energy Intensity of Iron and Steel Production in China and the U.S.

In this study, "energy intensity" is chosen as the index for comparison of the Chinese and U.S. iron and steel industries. It presents, within the prescribed boundary (as illustrated in Figure 3), the energy consumption per tonne of crude steel during production.

$$\text{Energy intensity} = \frac{\text{Energy Consumption of the iron and steel industry within the prescribed boundary}}{\text{Crude steel production within the prescribed boundary}}$$

The energy intensity of steel production is influenced by industry structure, technology, fuel choice, and materials--e.g., availability of scrap steel. The effects of these variables are isolated in this study's scenario analyses. Section 5.3 presents six scenarios to compare a range of effects within the steel industry of China and the U.S.

5.1. Energy Intensity of Iron and Steel Production in the U.S.

Final energy intensity (energy use per tonne of crude steel) for the U.S. iron and steel industry in 2006 is provided in Table 19. This value is calculated using the production data from Table 10 and the electricity and fuel consumption data from Table 15.

Crude steel production in the U.S. in 2006 was 98.2 Mt. In addition, there were 8.261 Mt of net imported ingots, blooms, billets, and slabs in 2006. Thus, total crude steel production used for the calculation of energy intensities in 2006 in this analysis was 106.461 Mt.

As explained in section 3.3.1 above, energy use for production of net imported ingots, blooms, billets, and slabs is calculated using international average conversion factors provided by the World Steel Association (worldsteel) since imported products can be from different countries and will thus vary in their energy consumption during production due to differences in production technology and energy structure (worldsteel, n.d.; worldsteel, 2008b).[14]

Under the base case scenario, the total electricity and fuel consumption in the iron and steel industry in the U.S. in 2006 based on the defined boundary of this study are 71,951 million kWh and 1,326,830 TJ, respectively. If these energy uses are divided by the production of crude steel given above, the electricity and fuel intensity can be calculated separately. The sum of the electricity and fuel intensity is given as the total final energy intensity.

Primary energy intensity is calculated by converting final electricity to primary energy intensity using the average power generation efficiency of fossil-fuel power plants in the U.S. as well as transmission and distribution losses.

By multiplying the final electricity intensity by the conversion factor given in Table 5 for the U.S., primary electricity intensity can be calculated. Finally, the fuel intensity (for which final and primary energy value is the same in this analysis) is then added to the primary electricity intensity to calculate the total primary energy intensity. Presenting the energy intensity in primary energy has the advantage of showing the relative efficiency of the power generation in both countries. Table 19 presents the various energy intensities explained above.

Table 19. Base Case - Energy Intensity of the U.S. Iron and Steel Industry in 2006

Scenario	Country	Electricity Intensity		Fuel Intensity		Final Energy Intensity[1]		Primary Energy Intensity with T&D [2]		Primary Energy Intensitywithout T&D [3]	
		kWh/t crude steel	kgce/t crude steel	GJ/t crude steel	kgce/t crude steel	GJ/t crude steel	kgce/t crude steel	GJ/t crude steel	kgce/t crude steel	GJ/t crude steel	kgce/t crude steel
Base	U.S.	675.84	83.44	12.46	425.25	14.90	508.69	19.98	681.68	19.47	665.71

[1] Total final energy intensity of the US iron and steel industry using the U.S. country-specific energy conversion factors for the purchased coke and auxiliary/intermediary products instead of worldsteel conversion factor would be 14.5 GJ/tonne crude steel which is around 2.7% less than the intensity calculated using worldsteel conversion factors.

[2] In final energy, electricity use is equal to the electricity consumption at the end-use. In primary energy with transmission and distribution losses (T&D), electricity use at the end-use is converted to the primary energy sources by taking into account the power generation efficiency (average net heat rate of power plants) and transmission and distribution losses.

[3] In primary energy without transmission and distribution losses (T&D), electricity use at the end-use is converted to the primary energy sources by taking into account *only* the power generation efficiency (average net heat rate of power plants). This is presented here because it is the common method in Chinese statistics. Thus, for consistency, both primary energy calculated using the international standard (with T&D losses) and using the Chinese standard (without T&D losses) are presented.

Conversions: GJ to kgce = 34.1208, kWh to kgce (final energy) = 0.1235.

Table 20. Base Case - Energy Intensity of the Chinese Iron and Steel Industry in 2006

Scenario	Country	Electricity Intensity		Fuel Intensity		Final Energy Intensity[1]		Primary Energy Intensity with T&D [2]		Primary Energy Intensity without T&D [2]	
		kWh/t crude steel	kgce/t crude steel	GJ/t crude steel	kgce/t crude steel	GJ/t crude steel	kgce/t crude steel	GJ/t crude steel	kgce/t crude steel	GJ/t crude steel	kgce/t crude steel
Base	China	431.66	53.02	21.54	734.96	23.11	788.53	26.30	897.38	25.97	886.12

[1] In final energy, electricity use is equal to the electricity consumption at the end-use. In primary energy with transmission and distribution losses (T&D), electricity use at the end-use is converted to the primary energy sources by taking into account the power generation efficiency (average net heat rate of power plants) and transmission and distribution losses.

[2] In primary energy without transmission and distribution losses (T&D), electricity use at the end-use is converted to the primary energy sources by taking into account *only* the power generation efficiency (average net heat rate of power plants).This is presented here because it is the common method in Chinese statistics. Thus, for consistency, both primary energy calculated using the international standard (with T&D losses) and using the Chinese standard (without T&D losses) are presented.

5.2. Energy Intensity of Iron and Steel Production in China

Table 20 shows the energy intensity (energy consumption per tonne crude steel) calculated based on the 2006 revised energy data for China, as presented in Section 4.2.2 above. Final energy intensity for the Chinese iron and steel industry in 2006 is provided in Table 20. This value is calculated using the Chinese production data from Table 11 and the electricity and fuel consumption data from Table 18. Crude steel production in China in 2006 was 421.02 Mt. In addition, there were 8.57 Mt of net exported ingots, blooms, billets, and slabs in 2006. Thus, total crude steel production used for the calculation of energy intensities in 2006 in this analysis was 412.45 Mt. Total electricity and fuel consumption of steel industry in China with embodied energy of net imported/exported auxiliary/intermediary products included are 178,039 GWh and 8,882,760 TJ, respectively. By dividing these energy consumption values by the production values, the energy intensities presented in Table 20 are obtained.

5.3. Scenario Analyses

In addition to the base case presented above, six variations to the base case were calculated to examine the impact of different assumptions on the iron and steel production energy intensity value for each country. The purpose of this scenario analysis is to determine which variables are most important for explaining energy intensity differences between China and the U.S.

The first scenario uses IEA typical fuel conversion factors (instead of country-specific fuel conversion factors used in the base case), country-specific electricity conversion factors, and worldsteel conversion factors for auxiliary/intermediary products. This scenario is intended to isolate the impact of the use of country-specific conversion factors on the overall comparative intensity.

The second scenario uses the country-specific fuel conversion factors, worldsteel electricity conversion factors for converting electricity from final to primary energy (9.8 MJ/kWh) (instead of country-specific electricity conversion factors used on the base case), and worldsteel conversion factors for auxiliary/intermediary products. This scenario is intended to analyze the impact on energy intensity caused by change of electric power conversion factor.

The third scenario uses the IEA typical fuel conversion factors (instead of country-specific fuel conversion factors used in the base case), worldsteel electricity conversion factors (9.8 MJ/kWh) ((instead of country-specific electricity conversion factors used on the base case), and the worldsteel conversion factors for auxiliary/intermediary products. The purpose of this scenario is to remove the effect of country-specific conversion factors and focus the intensity comparison on structural and efficiency effects.

The fourth scenario uses the country-specific fuel conversion factors, country-specific electricity conversion factors, worldsteel conversion factors for auxiliary/intermediary products, and China's EAF ratio in 2006 used for U.S. energy intensity calculation. This scenario is intended to analyze the impact on energy consumption caused by a change of the EAF ratio.

The fifth scenario uses the IEA typical fuel conversion factors, worldsteel electricity conversion factors (9.8MJ/kWh), worldsteel conversion factors for auxiliary/intermediary products, and China's EAF ratio in 2006 used for U.S. energy intensity calculation. This scenario is the same as the third scenario, but takes into account the impact on energy consumption caused by a change of EAF ratio.

Finally, the sixth scenario uses the country-specific fuel conversion factors, worldsteel conversion factors for auxiliary/intermediary products, and China's final to primary electricity conversion factor for U.S. energy intensity calculation. This scenario shows the impact of the different power generation efficiencies in each country on the primary energy intensity.

5.3.1. U.S. Iron and Steel Industry

Scenario 1: Energy intensity of each country's iron and steel industry using the IEA typical fuel conversion factors, country-specific electricity conversion factors, and worldsteel conversion factors for auxiliary/intermediary products

In order to develop a scenario in which a similar fuel conversion factor is used for both China and U.S. to understand the effect of differences in country-specific fuel conversion factors on the calculated energy intensities, the IEA typical fuel conversion factor is used for both countries. Table 21 provides the use of energy for all purposes (fuel and nonfuel) for the U.S. iron and steel industry in 2006 based on the use of energy in physical units given in Table 12 and the IEA's typical fuel conversion factors given in Table 5. The same methodology explained for the base case is used for the calculation of the energy intensity using the IEA typical fuel conversion factors for this scenario. The results of the analysis are shown in Table 22. As can be seen, for the U.S. iron and steel industry, the energy intensity using the IEA typical fuel conversion factors is almost the same as the intensities using the country-specific fuel conversion factors.

Scenario 2: Energy intensity of each country's iron and steel industry using country-specific fuel conversion factors, worldsteel electricity conversion factors for converting electricity from final to primary energy, and worldsteel conversion factors for auxiliary/intermediary products

In order to develop a scenario in which a similar conversion factor to convert electricity from final to primary energy is used for both China and U.S. to understand the effect of differences in country-specific electricity conversion factors on the calculated primary energy intensities, the worldsteel electricity conversion factor which is 9.8 MJ/kWh (0.334 kgce/kWh) is used for both countries (worldsteel, n.d.).

The results of the analysis are shown in Table 23. As can be seen, for the U.S. iron and steel industry, the primary energy intensity using the worldsteel electricity conversion factors is lower than that of using the U.S. country specific conversion factor for electricity. There is no change in electricity, fuel, and final energy intensities compared to the base case scenario since the change in this scenario was only the conversion factor that used to convert electricity from final to primary energy.

Scenario 3: Energy intensity of each country's iron and steel industry using IEA typical fuel conversion factors, worldsteel electricity conversion factors, and worldsteel conversion factors for auxiliary/intermediary products

This scenario is similar to scenario 2 with the difference that in this scenario the IEA typical fuel conversion factors are used instead of country-specific conversion factors. Therefore, the results of this scenario can be compared to scenario 2. The purpose of this scenario is to use a similar fuel conversion factor or both China and the U.S. while assessing the impact of the electricity conversion factor. The results of the analysis are shown in Table 24. As can be seen, the primary energy intensity of scenario 3 is lower than scenario 1 in which the IEA typical fuel conversion factor was used with the country-specific electricity conversion factor. Also, the results show that the primary energy intensities of scenario 2 and 3 are very close to each other.

Table 21. Energy Use for All Purposes (Fuel and Nonfuel) for the U.S. Iron and Steel Industry Using IEA Typical Fuel Conversion Factors, 2006 (TJ)

NAICS	Industry	Electricity	Residual Fuel Oil	Distillate Fuel Oil	Natural Gas	Coal used as Fuel[1]	Coal used for Coking[1]	Coke and Breeze	Other[2]	Shipments of Energy Sources Produced Onsite[3]	Total
331111	Iron and Steel Mills	184,313	18,997	6,034	346,276	17,887	398,496	224,074	41,147	145,568	1,091,655
3312	Steel Products from Purchased Steel	15,826	300	0	25,797	511	0.0	292	14,771	0	57,497

[1] Calculated from EIA, 2010b.

[2] Energy use of the category "Other" is based on IEA fuel conversion factors, since the energy use in physical units for this category was not available (only the first use of energy in energy units was available from U.S. DOE/EIA 2010b). Thus, the value used for this scenario is the same as for the base case. The share of energy use of the category "Other" is less than 4% of the total energy used; hence the impact of this assumption is minimal.

[3] For the category "Shipments of Energy Sources Produced Onsite", only the energy use value is provided by the EIA. However, this value is for the coke that is traded within the U.S iron and steel industry. Therefore, the ratio of the fuel conversion factor for coke from the IEA to fuel conversion factor for coke from the EIA was multiplied by the energy use of this category (in energy units) to determine the energy use of this category based on the IEA's fuel conversion factor for coke.

Table 22. Scenario 1 - Energy Intensity of the U.S. Iron and Steel Industry in 2006 Using IEA Typical Fuel Conversion Factors, Country - Specific Electricity Conversion Factors, and worldsteel Conversion Factors for Auxiliary/Intermediary Products

Scenario	Country	Electricity Intensity		Fuel Intensity		Final Energy Intensity		Primary Energy Intensity withT&D losses [1]		Primary Energy IntensitywithoutT&D losses [2]	
		kWh/t crude steel	kgce/t crude steel	GJ/t crude steel	kgce/t crude steel	GJ/t crude steel	kgce/t crude steel	GJ/t crude steel	kgce/t crude steel	GJ/t crude steel	kgce/t crude steel
1	U.S.	675.84	83.44	12.40	423.05	14.83	506.49	19.91	679.48	19.41	663.51

[1] In final energy, electricity use is equal to the electricity consumption at the end-use. In primary energy with transmission and distribution (T&D) losses, electricity use at the end-use is converted to the primary energy sources by taking into account the power generation efficiency (average net heat rate of power plants) and transmission and distribution losses

[2] In primary energy without transmission and distribution (T&D) losses, electricity use at the end-use is converted to the primary energy sources by taking into account *only* the power generation efficiency (average net heat rate of power plants). This is presented here because it is the common method in Chinese statistics. Thus, for consistency, both primary energy calculated using the international standard (with T&D losses) and using the Chinese standard (without T&D losses) are presented.

Conversions: GJ to kgce = 34.1208, kWh to kgce (final energy) = 0.1235.

Table 23. Scenario 2 - Energy Intensity of the U.S. Iron and Steel Industry in 2006 Using Country-Specific Fuel Conversion Factors, worldsteel Electricity Conversion Factors, and worldsteel Conversion Factors for Auxiliary/Intermediary Products

Scenario	Country	Electricity Intensity		Fuel Intensity		Final Energy Intensity		Primary Energy Intensity withT&D losses [1]		Primary Energy Intensity withoutT&D losses [2]	
		kWh/t crude steel	kgce/t crude steel	GJ/t crude steel	kgce/t crude steel	GJ/t crude steel	kgce/t crude steel	GJ/t crude steel	kgce/t crude steel	GJ/t crude steel	kgce/t crudesteel
2	U.S.	675.84	83.44	12.46	425.25	14.90	508.69			19.09	651.24

[1] In final energy, electricity use is equal to the electricity consumption at the end-use. In primary energy with T&D losses, electricity use at the end-use is converted to the primary energy sources by taking into account the power generation efficiency (average net heat rate of power plants) and transmission and distribution losses. Since the worldsteel electricity conversion factor used in Scenario 2 excludes T&D losses, the primary energy intensity with T&D losses based on worldsteel conversion factor is not available. [15]

[2] In primary energy without T&D losses, electricity use at the end-use is converted to the primary energy sources by taking into account *only* the power generation efficiency (average net heat rate of power plants). This is presented here because it is the common method in Chinese statistics. Thus, for consistency, both primary energy calculated using the international standard (with T&D losses) and using the Chinese standard (without T&D losses) are presented.

Conversions: GJ to kgce = 34.1208, kWh to kgce (final energy) = 0.1235.

Table 24. Scenario 3 - Energy Intensity of the U.S. Iron and Steel Industry in 2006 Using IEA Typical Fuel Conversion factors, worldsteel Electricity Conversion Factors, and worldsteel Conversion Factors for Auxiliary Products

Scenario	Country	Electricity Intensity		Fuel Intensity		Final Energy Intensity		Primary Energy Intensity withT&D losses [1]		Primary Energy Intensity withoutT&D losses [2]	
		kWh/t crude steel	kgce/t crude steel	GJ/t crude steel	kgce/t crude steel	GJ/t crude steel	kgce/t crude steel	GJ/t crude steel	kgce/t crude steel	GJ/t crude steel	kgce/t crudesteel
3	U.S.	675.84	83.44	12.40	423.05	14.83	506.49			19.02	649.04

[1] In final energy, electricity use is equal to the electricity consumption at the end-use. In primary energy with transmission and distribution (T&D) losses, electricity use at the end-use is converted to the primary energy sources by taking into account the power generation efficiency (average net heat rate of power plants) and transmission and distribution losses. Since the worldsteel electricity conversion factor used in Scenario 2 excludes T&D losses, the primary energy intensity with T&D losses based on worldsteel conversion factor is not available.

[2] In primary energy without transmission and distribution (T&D) losses, electricity use at the end-use is converted to the primary energy sources by taking into account *only* the power generation efficiency (average net heat rate of power plants). This is presented here because it is the common method in Chinese statistics. Thus, for consistency, both primary energy calculated using the international standard (with T&D losses) and using the Chinese standard (without T&D losses) are presented.

Conversions: GJ to kgce = 34.1208, kWh to kgce (in final energy) = 0.1235.

Scenario 4: Energy intensity of each country's iron and steel industry using country-specific fuel conversion factors, country-specific electricity conversion factors, worldsteel conversion factors for auxiliary/intermediary products, and China's EAF ratio in 2006 used for U.S. energy intensity calculation

A key factor influencing the energy intensity of a country's iron and steel industry is the share of steel that is produced by the more energy-intensive BF-BOF route compared to the share produced by the less energy-intensive EAF route. The iron and steel industry in the U.S. reflects the country's mature, industrialized economy that has an ample supply of scrap steel and that produced 56.9% of total steel by recycling this scrap steel in an EAF in 2006. China, on the other hand, is a developing country and due to scrap availability limitations only produced 10.5% of its total steel using the EAF route in 2006.[16] In order to account for this difference in the structure of the steel industry in China and the U.S., a scenario is presented in which the U.S. steel industry energy intensity is calculated using the share of EAF steel in China to illustrate how this structural difference changes the energy intensity value for the U.S. In order to calculate this scenario, the energy use and energy intensity by EAF and BF-BOF process routes is required. However, the EIA does not provide information on energy use by EAF and BF-BOF process routes for the U.S. Thus, an estimate of the breakdown of energy use by these two processes must be undertaken for the U.S.

The American Iron and Steel Institute (AISI) provides total final energy intensity for the production of one tonne of shipped steel by EAF technology and by BF-BOF technology in 2003 (Table 25) (AISI, 2005). To convert the energy intensities from units of per tonne of shipped steel to units of per tonne of crude steel in 2003, the ratio of the production of crude steel and shipped steel in 2003 from USGS (2004) is used (Table 25). These intensities are for the complete production line which includes casting, rolling and finishing for both EAF and BFBOF production lines. Also, these intensity values for EAF and BF-BOF production lines include the energy use for net imported auxiliary products such as the energy use for the production of pellet, lime, coke, and oxygen. However, these energy intensities do not include the embodied energy of the net imported pig iron, DRI, and ingots, blooms, billets, and slabs as well as the energy use for rolling and finishing of the net imported ingots, blooms, billets, and slabs to the U.S. iron and steel industry.

Table 25. Final Energy Intensity of the U.S Iron and Steel Industry in 2003

	Final Energy with auxiliary products energy use included except for embodied energy of imported pig iron, DRI, and ingot, slabs and billets (GJ/t shipped steel)	Final Energy with all auxiliary products energy use and embodied energy of imported pig iron, DRI, and ingot, slabs and billets included (GJ/t crude steel)
Final Energy Intensity of EAF production	6.1	7.5
Final Energy Intensity of BF-BOF production	22.7	25.3
Combined Final Energy Intensity of U.S steel industry*	14.3	16.2

Source: 2003 energy intensities: AISI, 2005.

* The combined final energy intensity is the total energy intensity of the iron and steel industry taking into account the share of EAF and BF-BOF production lines and their intensities.

The energy use for the production of net imported pig iron, DRI, and ingots, blooms, billets, and slabs is included in the calculation of the energy intensity from the MECS data for iron and steel industry. Hence, here also, this energy use must be included. It is explained above how the amount of crude steel required for the production of net imported ingots, blooms, billets, and slabs is calculated. Having this amount of crude steel and the energy intensity for the production of crude steel, the total energy use for the production of crude steel required for the production of net imported ingots, blooms, billets, and slabs can be calculated. Then, this total energy use can be divided by the crude steel production in 2006 to find out the energy intensity in GJ/tonne crude steel. The calculated intensities for the net imported ingots, blooms, billets, and slabs as well as the energy use given for the rolling and finishing processes of the net imported ingots, blooms, billets, and slabs (in MECS 2006, NAICS 3312) are then added to the energy intensities given in AISI (2005) for EAF and BF-BOF production lines in the U.S. The same approach is taken for net imported pig iron and DRI using the amount of imported material and their energy conversion factor from Table 6. At the end, the energy intensities that include the energy use of net imported auxiliary products (coke, lime, pellets, and oxygen) as well as pig iron, DRI, and ingots, blooms, billets, and slabs can be calculated.

Having the final energy intensities for each type of production line and the ratio of each type of production line (EAF and BF-BOF), the combined final energy intensity can be calculated using the following formula:

Combined EI = EI of EAF * EAF ratio + EI of BF-BOF * BF-BOF ratio

Where:

Combined EI: combined energy intensity of U.S steel industry (GJ/tonne **crude** steel)
EI of EAF: energy intensity of EAF production (GJ/tonne **crude** steel)
EI of BF-BOF: energy intensity of BF-BOF production (GJ/tonne **crude** steel)
EAF ratio: EAF ratio in U.S
BF-BOF ratio: BF-BOF ratio in U.S

The intensities in AISI (2005) are for 2003 and should be converted to 2006 values since this is the base year for this study. Table 26 provides the EAF ratio for the U.S. for 2003 and 2006 (worldsteel, 2009). Based on AISI (2005), it is assumed that the energy efficiency of the U.S. iron and steel industry improved at the rate of 0.7% per year from 2003 to 2006. This is used to calculate the 2006 intensity values from the 2003 values, as shown in Table 27.

Table 26. EAF Ratio in the U.S. in 2003 and 2006

	2003	2006
U.S. EAF ratio	51.0%	56.9%
U.S. BF -BOF ratio	49.0%	43.1%

Source: worldsteel, 2008.

In addition, to assign a share of EAF and BF-BOF production lines from the energy use of the net imported ingots, blooms, billets, and slabs, the share of EAF and BF-BOF in the steel production in the US in 2006 is used. These shares are used to divide the energy intensity of the imported ingots, blooms, billets, and slabs as well as the energy use given for rolling and finishing of them between EAF and BF-BOF steel production (see Table 27).

The 2006 final energy intensity presented earlier using the EIA's fuel conversion factors and worldsteel conversion factors for the net imported auxiliary and intermediary products is 14.9 GJ/tonne crude steel (Table 23). As can be seen in Table 27, the calculated combined final energy intensity for 2006 matches the value presented in Table 23. Therefore, the assumptions taken are reasonable since the energy intensity of the industry calculated from the EIA data and the AISI (2005) match.

Table 27. Final Energy Intensity of the U.S Iron and Steel Industry in 2006

	Final Energy with all auxiliary products energy use and embodied energy of imported pig iron, DRI, and ingot, slabs and billets included (GJ/t crude steel)
Final Energy Intensity of EAF production	7.3
Final Energy Intensity of BF-BOF production	24.8
Combined Final Energy Intensity of U.S steel industry*	14.9

Source: 2003 energy intensities: AISI, 2005.

* The combined final energy intensity is the total energy intensity of the iron and steel industry taking into account the share of EAF and BF-BOF production lines and their intensities. The formula for the calculation of the combined final energy intensity is given above.

Now that the energy intensities for the EAF and BF-BOF routes have been calculated for the U.S. iron and steel industry in 2006, Scenario 4 can be developed using the same calculation methodology but different EAF and BF-BOF ratios. In this scenario, the EAF and BF-BOF ratios of the Chinese steel industry in 2006 are used instead of the U.S ratios. In 2006, China produced 10.5% of its steel using the EAF route and 89.5% using the BF-BOF route.

To calculate the combined primary energy intensity from the calculated final energy intensity, the electricity and fuel intensity of each of the EAF and BF-BOF production routes must be calculated separately. Based on Stubbles (2000) it is assumed that around 45% of the final energy used in the EAF production line is electricity and the rest is fuel. Based on this share and the total final energy intensity of the EAF production line from Table 27 (7.3 GJ/tonne crude steel), the electricity intensity of the EAF production line is calculated to be 917.6 kWh/tonne crude steel. This value is in an acceptable range based on the Stubbles (2000) and Kirschen (2009). The fuel intensity is then calculated by subtracting the electricity intensity from the total final energy intensity.

Having the combined total electricity intensity of the U.S iron and steel industry from the base case scenario (Table 19) (675.8 kWh/tonne crude steel), the electricity intensity of the EAF production line from Table 25, and the EAF and BF-BOF ratios in U.S. in 2006, the electricity intensity of the BF-BOF production line can be calculated using the formula presented below. The fuel intensity of the BF-BOF production line can also be calculated using the same method.

EI BF-BOF = (EI U.S industry – EI EAF*EAF ratio)/BF-BOF ratio

Where:

EI BF-BOF: electricity intensity of the BF-BOF production line (kWh/tonne crude steel)
EI U.S industry: combined total electricity intensity of the U.S iron and steel industry from Table 19
EI EAF: electricity intensity of the EAF production line (kWh/tonne crude steel)

Table 28 presents the results for Scenario 4 showing that if the U.S. iron and steel industry had the same structure as the Chinese iron and steel industry in terms of the shares of EAF and BFBOF steel, the final energy intensity of U.S. steel production would be 22.96 GJ/tonne crude steel. This is a 54% increase in final energy intensity compared to the 14.9 GJ/tonne crude steel that was calculated for the base case in U.S. This is an important finding that shows that comparing only the total final energy intensity of the U.S and Chinese steel industry could be misleading and may not represent the real energy efficiency of the iron and steel industry in both countries since the difference in the EAF ratios of the two countries is large.

Scenario 5: Energy intensity of each country's iron and steel industry using the IEA typical fuel conversion factors, worldsteel electricity conversion factors, worldsteel conversion factors for auxiliary/intermediary products, and China's EAF ratio in 2006 used for U.S. energy intensity calculation.

The purpose of developing this scenario is to assess the impact of the share of EAF production while keeping all the conversion factors similar for both countries. The results of this scenario are presented in Table 29. As can be seen, the results of Scenario 5 are very close to the results of Scenario 4.

Table 28. Scenario 4 - Energy Intensity of the U.S. Iron and Steel Industry in 2006 Using Country-Specific Fuel Conversion Factors, Country-Specific Electricity Conversion Factors, worldsteel Conversion Factors for Auxiliary/Intermediary Products, and China's EAF Ratio in 2006 Used for U.S. Energy Intensity Calculation

Scenario	Country	Item	Electricity Intensity		Fuel Intensity		Total Final Energy Intensity		Primary Energy Intensity withT&D losses [1]		Primary Energy Intensity withoutT&D losses [2]	
			kWh/t crude steel	kgce/t crude steel	GJ/t crude steel	kgce/t crude steel	GJ/t crude steel	kgce/t crude steel	GJ/t crude steel	kgce/t crude steel	GJ/t crude steel	kgce/t crude steel
4	U.S.	Energy Intensity of EAF production line	917.64	112.72	4.04	137.77	7.34	250.48	14.24	485.94	13.56	462.61
		Energy Intensity of BF-BOF production line	356.63	43.81	23.51	802.02	24.79	845.83	27.47	937.34	27.21	928.27
		Combined Energy Intensity of U.S. steel industry [3]	**415.54**	**51.04**	**21.46**	**732.28**	**22.96**	**783.32**	**26.08**	**889.94**	**25.77**	**879.38**

[1] In final energy, electricity use is equal to the electricity consumption at the end-use. In primary energy with transmission and distribution (T&D) losses, electricity use at the end-use is converted to the primary energy sources by taking into account the power generation efficiency (average net heat rate of power plants) and transmission and distribution losses

[2] In primary energy without transmission and distribution (T&D) losses, electricity use at the end-use is converted to the primary energy sources by taking into account only the power generation efficiency (average net heat rate of power plants). This is presented here because it is the common method in Chinese statistics. Thus, for consistency, both primary energy calculated using the international standard (with T&D losses) and using the Chinese standard (without T&D losses) are presented.3 The combined final energy intensity is the total energy intensity of the iron and steel industry taking into account the share of EAF and BF-BOF production lines and their intensities. The formula for the calculation of the combined final energy intensity is given above.

Table 29. Scenario 5 - Energy Intensity of the U.S. Iron and Steel Industry in 2006 Using IEA Typical Fuel Conversion Factors, worldsteel Electricity Conversion Factors, worldsteel Conversion Factors for Auxiliary/Intermediary Products, and China's EAF Ratio in 2006 Used for U.S. Energy Intensity Calculation

Scenario	Country	Item	Electricity Intensity		Fuel Intensity		Total Final Energy Intensity		Primary Energy Intensity withT&D losses [1]		Primary Energy Intensity withoutT&D losses [2]	
			kWh/t crude steel	kgce/t crude steel	GJ/t crude steel	kgce/t crude steel	GJ/t crude steel	kgce/t crude steel	GJ/t crude steel	kgce/t crude steel	GJ/t crude steel	kgce/t crude steel
5	U.S.	Energy Intensity of EAF production line	926.25	113.78	4.08	139.06	7.41	252.84	N/A	N/A	13.15	448.78
		Energy Intensity of BF-BOF production line	345.26	42.41	23.63	806.17	24.87	848.58	N/A	N/A	27.01	921.62
		Combined Energy Intensity of U.S. steel industry [3]	**406.26**	**49.90**	**21.57**	**736.13**	**23.04**	**786.03**	N/A	N/A	**25.56**	**871.98**

[1] In final energy, electricity use is equal to the electricity consumption at the end-use. In primary energy with transmission and distribution (T&D) losses, electricity use at the end-use is converted to the primary energy sources by taking into account the power generation efficiency (average net heat rate of power plants) and transmission and distribution losses. Since the worldsteel electricity conversion factor used in Scenario 5 excludes T&D losses, the primary energy intensity with T&D losses based on worldsteel conversion factor is not available.[17]

[2] In primary energy without transmission and distribution (T&D) losses, electricity use at the end-use is converted to the primary energy sources by taking into account *only* the power generation efficiency (average net heat rate of power plants). This is presented here because it is the common method in Chinese statistics. Thus, for consistency, both primary energy calculated using the international standard (with T&D losses) and using the Chinese standard (without T&D losses) are presented.[3] The combined final energy intensity is the total energy intensity of the iron and steel industry taking into account the share of EAF and BF-BOF production lines and their intensities. The formula for the calculation of the combined final energy intensity is given above.

Conversions: GJ to kgce = 34.1208, kWh to kgce (final energy) = 0.1235.

Table 30. Scenario 6 - Energy Intensity of the U.S. Iron and Steel Industry Using Country-Specific Fuel Conversion Factors, worldsteel Conversion Factors for Auxiliary/Intermediary Products, and China's Final to Primary Electricity Conversion Factor for U.S. Energy Intensity Calculation

Scenario	Country	Electricity Intensity		Fuel Intensity		Final Energy Intensity		Primary Energy Intensity withT&D losses [1]		Primary Energy IntensitywithoutT&D losses [2]	
		kWh/t crude steel	kgce/t crude steel	GJ/t crude steel	kgce/t crude steel	GJ/t crude steel	kgce/t crude steel	GJ/t crude steel	kgce/t crude steel	GJ/t crude steel	kgce/t crude steel
6	U.S.	675.84	83.44	12.46	425.25	14.90	508.69	19.92	679.60	19.40	662.99

[1] In final energy, electricity use is equal to the electricity consumption at the end-use. In primary energy with transmission and distribution (T&D) losses, electricity use at the end-use is converted to the primary energy sources by taking into account the power generation efficiency (average net heat rate of power plants) and transmission and distribution losses.

[2] In primary energy without transmission and distribution (T&D) losses, electricity use at the end-use is converted to the primary energy sources by taking into account *only* the power generation efficiency (average net heat rate of power plants). This is presented here because it is the common method in Chinese statistics. Thus, for consistency, both primary energy calculated using the international standard (with T&D losses) and using the Chinese standard (without T&D losses) are presented.

Conversions: GJ to kgce = 34.1208, kWh to kgce (final energy) = 0.1235.

Table 31. Scenario 1 - Energy Intensity of China's Iron and Steel Industry in 2006 Using IEA Typical Fuel Conversion Factors, Country - Specific Electricity Conversion Factors, and worldsteel Conversion Factors for Intermediary Products

Scenario	Country	Electricity Intensity		Fuel Intensity		Final Energy Intensity		Primary Energy Intensity withT&D losses [1]		Primary Energy Intensity withoutT&D losses [2]	
		kWh/t crude steel	kgce/t crude steel	GJ/t crude steel	kgce/t crude steel	GJ/t crude steel	kgce/t crude steel	GJ/t crude steel	kgce/t crude steel	GJ/t crude steel	kgce/t crudesteel
1	China	429.43	53.09	21.01	719.27	22.65	769.87	25.75	878.49	25.41	867.18

[1] In final energy, electricity use is equal to the electricity consumption at the end-use. In primary energy with transmission and distribution (T&D) losses, electricity use at the end-use is converted to the primary energy sources by taking into account the power generation efficiency (average net heat rate of power plants) and transmission and distribution losses.

[2] In primary energy without transmission and distribution (T&D) losses, electricity use at the end-use is converted to the primary energy sources by taking into account *only* the power generation efficiency (average net heat rate of power plants). This is presented here because it is the common method in Chinese statistics. Thus, for consistency, both primary energy calculated using the international standard (with T&D losses) and using the Chinese standard (without T&D losses) are presented.

Conversions: GJ to kgce = 34.1208, kWh to kgce (final energy) = 0.1235.

Table 32. Scenario 2 - Energy Intensity of China's Iron and Steel Industry in 2006 Using Country-Specific Fuel Conversion Factors, worldsteel Electricity Conversion Factors, and worldsteel Conversion Factors for Auxiliary/Intermediary Products

Scenario	Country	Electricity Intensity		Fuel Intensity		Final Energy Intensity		Primary Energy Intensity withT&D losses [1]		Primary Energy Intensity withoutT&D losses [2]	
		kWh/t crude steel	kgce/t crude steel	GJ/t crude steel	kgce/t crude steel	GJ/t crude steel	kgce/t crude steel	GJ/t crude steel	kgce/t crude steel	GJ/t crude steel	kgce/t crudesteel
2	China	429.03	53.08	21.57	718.93	23.11	788.69	N/A	N/A	25.77	879.45

[1] In final energy, electricity use is equal to the electricity consumption at the end-use. In primary energy with transmission and distribution (T&D) losses, electricity use at the end-use is converted to the primary energy sources by taking into account the power generation efficiency (average net heat rate of power plants) and transmission and distribution losses. Since the worldsteel electricity conversion factor used in Scenario 2 excludes T&D losses, the primary energy intensity with T&D losses based on worldsteel conversion factor is not available.[18]

[2] In primary energy without transmission and distribution (T&D) losses, electricity use at the end-use is converted to the primary energy sources by taking into account *only* the power generation efficiency (average net heat rate of power plants). This is presented here because it is the common method in Chinese statistics. Thus, for consistency, both primary energy calculated using the international standard (with T&D losses) and using the Chinese standard (without T&D losses) are presented.

Conversions: GJ to kgce = 34.1208, kWh to kgce (final energy) = 0.1235.

Table 33. Scenario 3 - Energy Intensity of China's Iron and Steel Industry in 2006 Using IEA Typical Fuel Conversion factors, worldsteel Electricity Conversion Factors, and worldsteel Conversion Factors for Auxiliary/Intermediary Products

Scenario	Country	Electricity Intensity		Fuel Intensity		Final Energy Intensity		Primary Energy Intensity withT&D losses [1]		Primary Energy IntensitywithoutT&D losses [2]	
		kWh/t crude steel	kgce/t crude steel	GJ/t crude steel	kgce/t crude steel	GJ/t crude steel	kgce/t crude steel	GJ/t crude steel	kgce/t crude steel	GJ/t crude steel	kgce/t crude steel
3	China	430.65	53.09	21.01	719.27	22.65	769.87	N/A	N/A	25.23	860.88

[1] In final energy, electricity use is equal to the electricity consumption at the end-use. In primary energy with transmission and distribution (T&D) losses, electricity use at the end-use is converted to the primary energy sources by taking into account the power generation efficiency (average net heat rate of power plants) and transmission and distribution losses. Since the worldsteel electricity conversion factor used in Scenario 3 excludes T&D losses, the primary energy intensity with T&D losses based on worldsteel conversion factor is not available.[19]

[2] In primary energy without transmission and distribution (T&D) losses, electricity use at the end-use is converted to the primary energy sources by taking into account *only* the power generation efficiency (average net heat rate of power plants). This is presented here because it is the common method in Chinese statistics. Thus, for consistency, both primary energy calculated using the international standard (with T&D losses) and using the Chinese standard (without T&D losses) are presented.

Scenario 6: Energy intensity of each country's iron and steel industry using the country-specific fuel conversion factors, worldsteel conversion factors for auxiliary/intermediary products, and China's final to primary electricity conversion factor for U.S. energy intensity calculation

Another key difference between countries that has an effect on the primary energy consumption level and hence the primary energy intensity is the efficiency of electricity generation, transmission, and distribution in each country. A sixth scenario is constructed to illustrate the impact of using the Chinese electricity conversion factor to calculate primary energy use of the U.S. iron and steel industry.

In this scenario, the electricity, fuel, and final energy intensities are the same as the base case. However, to convert the final energy intensities to primary energy intensities for the U.S. iron and steel industry, the Chinese electricity conversion factor of 0.376 kgce/kWh from Table 9 is used instead of the U.S. electricity conversion factor of 0.379 kgce/kWh. This scenario is used to highlight the effect of power generation efficiency and T&D losses. The electricity conversion factor for China is almost same as the one in the U.S.; therefore, the total primary energy intensity of the U.S. iron and steel industry using the Chinese electricity conversion factor is about the same as the actual value using the U.S electricity conversion factor (base case scenario). The result for Scenario 6 is presented in Table 30.

5.3.2. Chinese Iron and Steel Industry

Scenario 1: Energy intensity of each country's iron and steel industry using the IEA typical fuel conversion factors, country-specific electricity conversion factors, and worldsteel conversion factors for auxiliary/intermediary products

As discussed in Section 3.3.1, the IEA typical coefficient for fuel is for dry-based fuel, but energy consumption data published by the *China Energy Statistical Yearbook* is wet-based. Thus, under this scenario, the Chinese coal consumption data is converted to dry-based coal use, based on a 10% moisture content conversion factor. In addition, the results from this scenario can be used to analyze the impact of fuel conversion coefficient on China's energy consumption in iron and steel industry.

The analysis shows that if the IEA typical energy conversion factors are used to calculate the energy consumption of China's iron and steel industry, the energy consumption values are slightly lower (about 2%) than the resulting value when using China's country-specific energy conversion factors (see Table 31).

Scenario 2: Energy intensity of each country's iron and steel industry using country-specific fuel conversion factors, worldsteel electricity conversion factors for converting electricity from final to primary energy, and worldsteel conversion factors for auxiliary/intermediary products

The results of this scenario are presented in Table 32. This scenario can be used to analyze the impact of electric power conversion coefficient on China's energy consumption in iron and steel industry. The energy consumption per tonne of crude steel calculated based on the electric power conversion coefficient from worldsteel shows no change in final energy intensity as expected but reduces the primary energy intensity by 0.75% compared to the base case. This is the result of electric power equivalent coefficient at 9.8 MJ/kWh (0.334 kgce/kWh), which is lower than China country-specific coefficient of 10.26 MJ/kWh (0.3505 kgce/kWh).

Scenario 3: Energy intensity of each country's iron and steel industry using IEA typical fuel conversion factors, worldsteel electricity conversion factors, and worldsteel conversion factors for auxiliary/intermediary products

It needs to be noted that the analysis conducted in Scenarios 1 and 2 can describe the impact on the energy consumption of China's iron and steel industry as a result of the changes of conversion coefficient for fuel and electricity. In other words, the results can only be used to analyze changes in energy consumption within each individual country, not between countries. The comparison between countries can be made possible only when the conversion coefficient for fuel, electricity, and net imported auxiliary products power is set the same. Accordingly, the analysis of Scenario 3 serves as a foundation for comparison between countries (Table 33).

Scenario 4: Energy intensity of each country's iron and steel industry using country-specific fuel conversion factors, country-specific electricity conversion factors, worldsteel conversion factors for auxiliary/intermediary products, and US EAF ratio in 2006 used for China's energy intensity calculation

China's iron and steel industry is at different development stage from the U.S. steel industry. With production growth remaining upward, scrap steel is still in short supply. The ratio of EAF steel production in total production in China has not increased, but instead has declined in recent years due to the rapid growth of BF-BOF steel production. However, China's iron and steel industry has been working to eliminate inefficient facilities, to restructure, and has been actively adopting energy-efficient techniques and technology. These actions have led to the downward trend of the industry's unit energy consumption. In other words, China's steel industry has improved energy efficiency through measures other than increased EAF production. Given China's ongoing economic growth, demand is likely to remain robust, limiting the availability of scrap steel and the ability to produce more EAF steel.

The increase in the EAF steel production ratio leads to the decline in the BOF steel production ratio from total steel produced. This also corresponds to the decline in the blast furnace production ratio in the iron-making process. The formula for calculation of energy consumption by the industry is as follows:

Energy consumption per ton steel = energy consumption of the iron-making process × iron to steel ratio + energy consumption of BOF steel-making process × BOF steel ratio + energy consumption of EAF steel-making process × EAF steel ratio + energy consumption of rolling steel-making process + others

$$\text{Iron to steel ratio} = \frac{\text{raw iron used by BOF} \times \text{BOF steel production} + \text{raw iron used by EAF} \times \text{EAF steel production} + \text{export raw iron}}{\text{crude steel production}}$$

The energy consumption of individual processes in 2006 is shown in Table 34.

Table 34. Energy Consumption of Individual Iron and Steel Industry Processes in China, 2006

Process	Coking	Sintering	Iron-making	BOF	EAF
2006 Process Energy Consumption (kgce/t)	126.12	54.65	436.67	11.49	105.6

The iron to steel ratio of China's iron and steel industry in 2006 is 0.98 (Editorial Board of China Iron and Steel Industry Yearbook, 2007). China's EAF steel ratio is 10.5% while BOF steel ratio reached 89.5% of total steel production in 2006. Accordingly, the ratio of exported raw iron to crude steel production is 0.0017. Raw iron used by BOF is 1.054~1.061t/t. In 2006, the relation between iron to steel ratio, EAF steel ratio and BOF steel ratio for the industry is as follows:

Iron to steel ratio= (1.054~1.061) × BOF steel ratio + (0.275~0.33) ×EAF steel ratio + 0.0017

Other conditions being equal, with 1% increase in EAF steel ratio, the iron to steel ratio decreases by 0.0073~0.0080. Based on calorific coefficient, the energy consumption in the process before iron-making in 2006 is 572.03 kgce/t. With 1% increase in EAF steel ratio, the energy consumption per ton steel decreases 3.26 ~ 3.61 kgce/t steel, with an average of 3.43 kgce/t. Based on equivalent coefficient, with 1% increase in EAF steel ratio, the energy consumption per ton steel decreases by 2.64 ~ 3.01 kgce/t steel, with an average of 2.83 kgce/t.

Table 35 shows the energy consumption of China's iron and steel industry under Scenario 4 when China's EAF steel ratio reaches the US level of 56.9% in 2006, which is calculated based on the above information. According to this table, when China's EAF steel ratio reaches the US level of 56.9% in 2006, the final energy intensity of China's iron and steel industry decreases significantly by 20.2%, and the primary energy intensity drops by 14.8% compared to the base case. Thus, the ratio of EAF has significant impact on energy consumption per ton of steel.

Scenario 5: Energy intensity of each country's iron and steel industry using the IEA typical fuel conversion factors, worldsteel electricity conversion factors, worldsteel conversion factors for auxiliary/intermediary products, and U.S. EAF ratio in 2006 used for China's energy intensity calculation

The analysis of energy consumption of per ton steel in Scenario 5 is based on IEA fuel conversion coefficients and the worldsteel electric power conversion coefficient, with both coefficients close to the international level. The fuel conversion coefficients used by each country will be different due to variances in fuel quality. Similarly, electricity conversion coefficient from final to primary energy will be different due to different development stage of power generation of each country. It is thus fair to use the world average for international comparisons. The results of this analysis are presented in Table 36. The analysis result of Scenario 5 is based on the world average conversion factors, while taking into account the difference in the EAF steel ratio.

Table 35. Scenario 4 - Energy Intensity of the China's Iron and Steel Industry in 2006 Using Country-Specific Fuel Conversion Factors, Country-Specific Electricity Conversion Factors, worldsteel Conversion Factors for Auxiliary/Intermediary Products, and the U.S. EAF Ratio in 2006 Used for China's Energy Intensity Calculation

Scenario	Country	Electricity Intensity		Fuel Intensity		Final Energy Intensity		Primary Energy Intensity withT&D losses [1]		Primary Energy Intensity withoutT&D losses [2]	
		kWh/t crude steel	kgce/t crude steel	GJ/t crude steel	kgce/t crude steel	GJ/t crude steel	kgce/t crude steel	GJ/t crude steel	kgce/t crude steel	GJ/t crude steel	kgce/t crudesteel
4	China	552.55	67.87	16.45	561.29	18.44	629.19	22.54	769.24	22.12	754.75

[1] In final energy, electricity use is equal to the electricity consumption at the end-use. In primary energy with transmission and distribution (T&D) losses, electricity use at the end-use is converted to the primary energy sources by taking into account the power generation efficiency (average net heat rate of power plants) and transmission and distribution losses.

[2] In primary energy without transmission and distribution (T&D) losses, electricity use at the end-use is converted to the primary energy sources by taking into account *only* the power generation efficiency (average net heat rate of power plants). This is presented here because it is the common method in Chinese statistics. Thus, for consistency, both primary energy calculated using the international standard (with T&D losses) and using the Chinese standard (without T&D losses) are presented.

Conversions: GJ to kgce = 34.1208, kWh to kgce (final energy) = 0.1235.

Table 36. 2006 Scenario 5 - Energy Intensity of the China's Iron and Steel Industry in 2006 Using IEA Typical Fuel Conversion Factors, worldsteel Electricity Conversion Factors, worldsteel Conversion Factors for Auxiliary/Intermediary Products, and the U.S. EAF Ratio in 2006 Used for China's Energy Intensity Calculation

Scenario	Country	Electricity Intensity		Fuel Intensity		Final Energy Intensity		Primary Energy Intensity withT&D losses [1]		Primary Energy Intensity withoutT&D losses [2]	
		kWh/t crude steel	kgce/t crude steel	GJ/t crude steel	kgce/t crude steel	GJ/t crude steel	kgce/t crude steel	GJ/t crude steel	kgce/t crude steel	GJ/t crude steel	kgce/t crude steel
5	China	561.29	68.95	15.88	541.84	17.90	610.76	N/A	N/A	21.38	729.50

[1] In final energy, electricity use is equal to the electricity consumption at the end-use. In primary energy with transmission and distribution (T&D) losses, electricity use at the end-use is converted to the primary energy sources by taking into account the power generation efficiency (average net heat rate of power plants) and transmission and distribution losses. Since the worldsteel electricity conversion factor used in Scenario 5 excludes T&D losses, the primary energy intensity with T&D losses based on worldsteel conversion factor is not available.[20]

[2] In primary energy without transmission and distribution (T&D) losses, electricity use at the end-use is converted to the primary energy sources by taking into account *only* the power generation efficiency (average net heat rate of power plants). This is presented here because it is the common method in Chinese statistics. Thus, for consistency, both primary energy calculated using the international standard (with T&D losses) and using the Chinese standard (without T&D losses) are presented.

Table 37. Scenario 6 - Energy Intensity of the China's Iron and Steel Industry Using Country-Specific Fuel Conversion Factors, worldsteel Conversion Factors for Auxiliary/Intermediary Products, and the U.S. Final to Primary Electricity Conversion Factor for China's Energy Intensity Calculation

Scenario	Country	Electricity Intensity		Fuel Intensity		Final Energy Intensity		Primary Energy Intensity withT&D losses [1]		Primary Energy IntensitywithoutT&D losses [2]	
		kWh/t crude steel	kgce/t crude steel	GJ/t crude steel	kgce/t crude steel	GJ/t crude steel	kgce/t crude steel	GJ/t crude steel	kgce/t crude steel	GJ/t crude steel	kgce/t crude steel
6	China	428.36	52.62	21.57	735.99	23.11	788.53	26.33	898.51	26.01	887.48

[1] In final energy, electricity use is equal to the electricity consumption at the end-use. In primary energy with transmission and distribution (T&D) losses, electricity use at the end-use is converted to the primary energy sources by taking into account the power generation efficiency (average net heat rate of power plants) and transmission and distribution losses.

[2] In primary energy without transmission and distribution (T&D) losses, electricity use at the end-use is converted to the primary energy sources by taking into account *only* the power generation efficiency (average net heat rate of power plants).

This is presented here because it is the common method in Chinese statistics. Thus, for consistency, both primary energy calculated using the international standard (with T&D losses) and using the Chinese standard (without T&D losses) are presented.

Conversions: GJ to kgce = 34.1208, kWh to kgce (final energy) = 0.1235.

Table 38. Energy Intensity for the Iron and Steel Industry in China and the U.S. (2006)

No.	Scenarios	Country	Final Energy Intensity		Primary Energy Intensity*	
			GJ/t	kgce/t	GJ/t	kgce/t
Base	Country-specificfuel conversion factors Country-specific electricity conversion factors worldsteel conversion factors aux/intermediary products	U.S.	14.90	508.69	19.98	681.68
		China	23.11	788.53	26.30	897.29
1	IEA typicalfuel conversion factors Country-specific electricity conversion factorsworldsteel conversion factors aux/intermediary products	U.S.	14.83	506.49	19.91	679.48
		China	22.56	769.87	25.75	878.49
2	Country-specific fuel conversion factors worldsteel electricity conversion factors worldsteel conversion factors aux/intermediary products	U.S.	14.90	508.69	19.09	651.24
		China	23.11	788.69	25.77	879.45
3	IEA typicalfuel conversion factors worldsteel electricity conversion factorsworldsteel conversion factors aux/intermediary products	U.S.	14.83	506.49	19.02	649.04
		China	22.65	769.87	25.23	860.88
4a	Country-specificfuel conversion factors Country-specific electricity conversion factors worldsteel conversion factors aux/intermediary products China 2006 EAF ratio used for U.S.	U.S.	22.96	783.32	26.08	889.94
	(Base Scenario) Country-specific fuel conversion factors Country-specific electricity conversion factors worldsteel conversion factors aux/intermediary products	China	23.11	788.53	26.30	897.29
4b	(Base Scenario) Country-specific fuel conversion factors Country-specific electricity conversion factors worldsteel conversion factors aux/intermediary products	U.S.	14.90	508.69	19.98	681.68
	Country-specific fuel conversion factors Country-specific electricity conversion factors worldsteel conversion factors aux/intermediary products U.S. 2006 EAF ratio used for China	China	18.44	629.19	22.54	769.24

No.	Scenarios	Country	Final Energy Intensity		Primary Energy Intensity*	
			GJ/t	kgce/t	GJ/t	kgce/t
5a	IEA typicalfuel conversion factors worldsteel electricity conversion factors worldsteel conversion factors aux/intermediary products China 2006 EAF ratio used for U.S.	U.S.	23.04	786.03	26.09	890.27
	(Scenario 3) IEA typical fuel conversion factors worldsteel electricity conversion factorsworldsteel conversion factors aux/intermediary products	China	22.65	769.87	25.23	860.88
5b	(Scenario 3) IEA typical fuel conversion factors worldsteel electricity conversion factors worldsteel conversion factors aux/intermediary products	U.S.	14.83	506.49	19.02	649.04
	IEA typical fuel conversion factors worldsteel electricity conversion factors worldsteel conversion factors aux/intermediary products U.S. 2006 EAF ratio used for China	China	17.90	610.76	21.38	729.50
6	Country-specific fuel conversion factors worldsteel conversion factors aux/intermediary products China final to primary electricity conversion factor for U.S. energy intensity calculation	U.S.	14.90	508.69	19.92	679.60
	Country-specific fuel conversion factors worldsteel conversion factors aux/intermediary products U.S. final to primary electricity conversion factor for China energy intensity calculation	China	23.11	788.53	26.33	898.51

* For the base Scenario and Scenarios 1, 4 and 6 the primary energy value includes T&D losses, whereas for Scenarios 2, 3, and 5 the primary energy value is calculated based on the worldsteel conversion factor which excludes T&D losses.

** Values are in metric tons of crude steel.

Scenario 6: Energy intensity of each country's iron and steel industry using the country-specific fuel conversion factors, worldsteel conversion factors for auxiliary/intermediary products, and the U.S. final to primary electricity conversion factor for China's energy intensity calculation.

The results of this scenario are only applicable to the comparison analysis of power generation efficiencies of different countries. In this scenario, the electricity, fuel, and final energy intensities are the same as the base case. However, to convert the final energy intensities to primary energy intensities for the China iron and steel industry, the U.S. electricity conversion factor from Table 7 is used instead of the Chinese electricity conversion factor.

5.3.3. Summary of Scenario Results

Table 38 and Figure 6 present the results of the base case and the six scenarios, providing information on the calculated electricity intensity, fuel intensity, final energy intensity, and primary energy intensity for the U.S. and China.

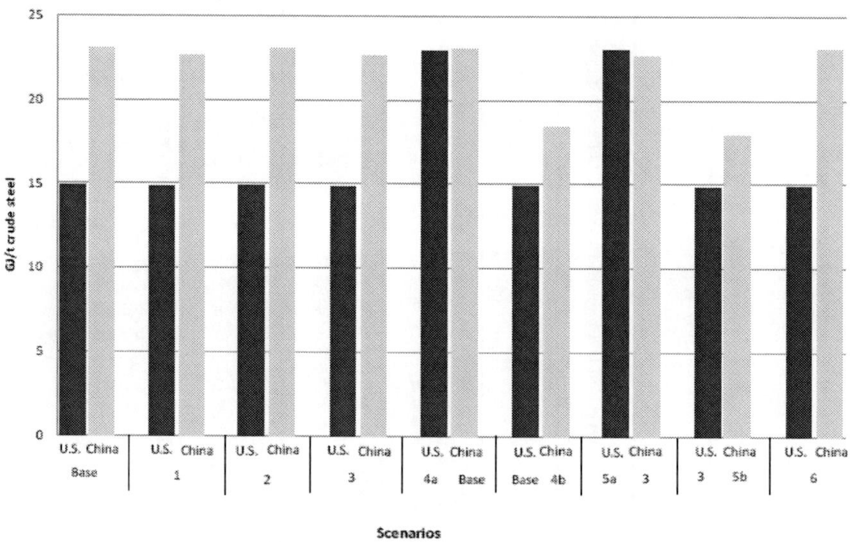

Note: For the base Scenario and Scenarios 1, 4 and 6 the primary energy value includes T&D losses, whereas for Scenarios 2, 3, and 5 the primary energy value is calculated based on the worldsteel conversion factor which excludes T&D losses.

Figure 6. Comparison of Energy Intensity for the Iron and Steel Industry in China and the U.S.: Base Case and Scenarios (2006).

5.3.4. Uncertainty and Unresolved Issues

- **Fuel consumption data for China's steel industry**: In Table 16 *Final Energy Consumption (Physical Amount) of Smelting & Pressing of Ferrous Metals (including ferroalloys) in 2006,* energy consumption in physical values is provided from the 2009 China Energy Statistical Yearbook. However, these values include energy use for ferroalloys which should be deducted for this analysis since ferroalloys are not within the boundary of this study. The energy use of ferroalloys industry is based on statistics provided to the China Iron and Steel Research Institute (CISRI) by China's National Bureau of Statistics (NBS). The energy use by China's ferrous alloy industry equal is to 1,142,704 TJ. This energy use is deducted from the total energy use shown in Table 16. This value has been confirmed by Chinese experts from the ferrous alloy industry, but there is not a published reference for this value.
- **The IEA fuel conversion factors for other bituminous coal and clean coal**: The country-specific fuel conversion factors for these coals used in the U.S. are for dry base coal, but CISRI has explained that the country-specific coal consumption data in China are for wet base coal. Thus, for the calculation of the Scenarios in which the IEA conversion factors are used, China's reported wet base coal value was multiplied by 0.9 to convert to dry base coal. As a result, the overall coal energy use in China is reduced as well as the resulting energy intensity. The choice of 0.9 depends on the moisture content of the coal in China compared to that used by IEA for defining the fuel conversion factor. There is uncertainty about this value which adds minor uncertainty to the results for scenarios that use the IEA fuel conversion factors.
- **Results of Scenario 5a**: The factor that contributes to the result of Scenario 5a where the U.S. steel intensity is slightly higher than the Chinese steel intensity is that this scenario normalizes the comparison by using the IEA typical fuel conversion factors instead of country-specific fuel conversion factors. The main difference between the actual, country-specific energy conversion factors in the U.S. and China is for other bituminous coal and clean coal. As discussed above, the country-specific fuel conversion factors for these coals used in the U.S. are for dry base coal, but the country-specific fuel conversion factors used in China are for wet base coal. For the calculation of Scenario 5a, however, China's reported wet base coal value was

multiplied by 0.9 to convert to dry base coal. As a result, the overall coal energy use in China is reduced as well as the resulting energy intensity. Secondly, the amount of pig iron used in the EAF also varies between the two countries.

- **Breakdown of energy use of Chinese steel industry by BOF and EAF production lines**: The energy intensities of the EAF and BF-BOF production lines in China could not be calculated separately from China's scenario 4b results. This was because it is very difficult to separate the EAF production route and BF-BOF production route in China, as the situation in China is very different from the U.S. Unlike in the U.S., in China the EAF production route and BF-BOF production route are often included in an integrated plant, which also means that they share the auxiliary equipment and materials. The following analysis has been undertaken to indirectly calculate the energy intensities of the EAF and BF-BOF production lines in China separately:

The energy intensities calculated for China in the base year (using the Chinese EAF ratio) and the energy intensities calculated for China in scenario 4b (using US EAF ratio) were used. Having the EAF ratios in China and US and the intensities from these two scenarios, the energy intensities of EAF and BF-BOF can be calculated separately from the following equation:

1. Chinese EAF final energy intensity * EAF ratio in China + Chinese BF-BOF final energy intensity * BF-BOF ratio in China = Final Energy intensity in base case
2. Chinese EAF final energy intensity * EAF ratio in US + Chinese BF-BOF final energy intensity * BF-BOF ratio in US = Final Energy intensity in scenario 4b

Then:

1. Chinese EAF final energy intensity * 10.5% + Chinese BF-BOF final energy intensity * 89.5% = 23.11 (GJ/t crude steel)
2. Chinese EAF final energy intensity * 56.9% + Chinese BF-BOF final energy intensity * 43.1% = 18.44 (GJ/t crude steel)

By solving these two equations with two unknown variables, the Chinese EAF final energy intensity and Chinese BF-BOF final energy intensity can be calculated. The results are:

- Chinese EAF final energy intensity = 14.11 (GJ/t crude steel)
- Chinese BF-BOF final energy intensity = 24.17 (GJ/t crude steel)

Table 39. Comparison of calculated final energy intensities for EAF and BF-BOF production lines in China and the U.S in 2006

	Final energy intensity (GJ/t crude steel)		Primary energy intensity (GJ/t crude steel)	
	China	U.S.	China	U.S.
EAF production line	14.11	7.34	19.01	14.24
BF-BOF production line	24.17	24.79	27.17	27.47

It can be seen that the final energy intensity for EAF steel production in China is significantly higher than that in the U.S., while the intensity of BF-BOF steel production in China is slightly lower than that of in the U.S. One reason for this could be the higher share of pig iron used in EAFs in China compared to the U.S. In the U.S., about 10% of the feed to the EAF is pig iron which is produced in a blast furnace and is highly fuel intensive, while about 28% of the feed to the EAF in China is pig iron. The higher the share of pig iron, the lower the electricity use in EAF, yet the higher the fuel used for the production of crude steel by EAF production line. However, the significantly higher intensity for EAF in China might not be all explained by the higher share of pig iron in the charge mix and requires further investigation. However, the difference in primary energy intensity of EAF is not as significant as the final energy intensity value. The reason for this is that in U.S. the EAF production lines use more scrap as input and less pig iron and thus use more electricity which is presented in final energy. When converting electricity from final to primary energy, the gap between energy intensity of EAF production lines in two countries decreases. It would be best if the Chinese could calculate directly the EAF and BF-BOF energy intensities separately from the energy used in those production lines as it is done for the U.S. in scenario 4a of this report.

- **Values for the energy intensity of EAF and BF-BOF production lines in U.S.** were obtained from AISI (2005). The intensities given in

AISI (2005) were for the year 2003. However, the base year for this study is 2006. Thus, a 0.7% improvement in energy efficiency per year was assumed which was also obtained from AISI (2005) and applied to the 2003 intensities in order to get the 2006 energy intensities. The other issues with AISI intensities for EAF and BF-BOF used is that it was assumed for this study that these intensities include the energy use for the production of auxiliary products but exclude the embodied energy of imported ingot, slabs and billets, DRI, and pig iron. These assumptions could also add uncertainty to the results and should be verified.

- **Share of energy use for electricity and fuel**: As mentioned in the calculation of scenario 4a for the U.S., it was assumed that 45% of the final energy use in a EAF production line (as defined in this study) is electricity and the rest is fuel, while only 5% of the final energy use in BF-BOF production line (as defined) is electricity and 95% is fuel. These shares could also vary slightly from one plant to another and may add uncertainty to the results.

5.4. Explanatory Variables

The purpose of the analysis presented in this report is to develop and test a methodology for quantifying and comparing the energy intensity of steel production in China and the U.S. with defined boundaries and conversion factors. This section provides a discussion of some possible reasons that the energy intensity values differ in the two countries. Six explanatory variables are discussed: 1) the age of steel manufacturing facilities in each country, 2) the share of EAF steel in total steel production, 3) the level of penetration of energy-efficient technologies, 4) the scale of production equipment, 5) fuel shares in the iron and steel industry, and 6) the final steel product mix in both countries.

5.4.1. Age of Steel Manufacturing Facilities

As is evident from Figure 1, most of China's steel production capacity has been constructed since 2000, when annual production jumped from 129 Mt to 627 Mt in 2010. During that same time, production in the U.S. dropped from 102 Mt to 80 Mt.

While there are no data available on the exact age of each steel enterprise in China, we can infer from the production data that in 2011, about 500 Mt of production (or about 80%) is from plants that are 10 years old or younger. In contrast, the average age of BOF vessels in the U.S. is 31.5 years (AIST, 2010a) and the average age of EAF furnaces in the U.S. is 30.9 years (AIST, 2010b). Even though the vessels have been relined and other upgrades have been made to the U.S. facilities, they are overall older than most of the steel production facilities in China. However, it should also be noted that not all of the new Chinese plants have necessarily installed the most energy-efficient technologies.

5.4.2. Structure of the Steel Manufacturing Sector

The structure of the steel manufacturing sector is one of the key variables that explains the difference in energy intensity values in China and the U.S. since EAF steel production uses significantly less energy for the production of one tonne of steel. In 2006, the share of EAF steel production in total steel production was 10.5% in China and 56.9% in the U.S. The world average EAF production in 2006 was 31.6% (see Figure 7).

Scenarios 4 and 5 calculate the total U.S. energy intensity using the share of EAFs in China and the total Chinese energy intensity using the share of EAFs in the U.S., respectively.

Scenario 4a, which relies on country-specific fuel and electricity conversion factors and worldsteel conversion factors for auxiliary and intermediate products, found that if the U.S. iron and steel industry had the same structure as the Chinese iron and steel industry in terms of the shares of EAF steel, the final energy intensity of U.S. steel production would be 22.96 GJ/tonne crude steel using the same conversion factors. This value should be compared to the Base Scenario for China which resulted in a final energy intensity of 23.11 GJ/tonne crude steel using the same conversion factors. Conversely, Scenario 4b shows that if the Chinese steel industry had the same structure as the U.S. steel industry in terms of shares of EAF steel, the final energy intensity of Chinese steel production would be 18.44 GJ/tonne crude steel using the same conversion factors. This value should be compared to the Base Scenario for the U.S which resulted in a final energy intensity of 14.9 GJ/tonne crude steel.

Scenario 5a, which uses IEA typical fuel conversion factors and v conversion factors for electricity and auxiliary and intermediate products, found that if the U.S. iron and steel industry had the same structure as the Chinese iron and steel industry in terms of the shares of EAF steel, the final energy intensity of U.S. steel production would be 23.04 GJ/tonne crude steel. This value should be compared to Scenario 3 for China, in which the resulting steel energy intensity was calculated to be 22.65 GJ/tonne crude steel using the same conversion factors. Conversely, Scenario 5b shows that if the Chinese steel industry had the same structure as the U.S. steel industry in terms of shares of EAF steel, the final energy intensity of Chinese steel production would be 17.9 GJ/tonne crude steel using the same conversion factors. This value should be compared to Scenario 3 for the U.S which resulted in a final energy intensity of 14.83 GJ/tonne crude steel using the same conversion factors.

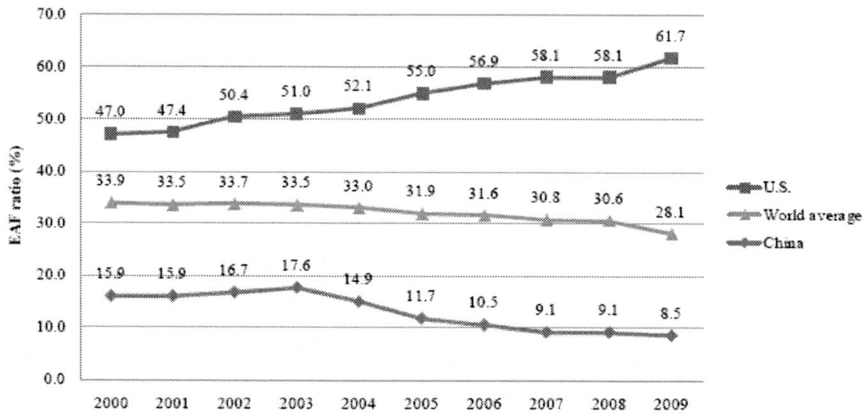

Source: World Steel Association, 2009.

Figure 7. Share of EAF in Total Steel Production in China and the U.S. and World Average Values.

5.4.2. Fuel Shares

The share of different fuels used in the iron and steel industry in both countries is also another important variable that should be considered. The fuel shares will influence the energy intensity of the iron and steel industry, as well as the related carbon dioxide emissions. Figures 8 and 9 show the shares of different fuels used (both as fuel and nonfuel) in the U.S. and Chinese iron and steel industries. As can be seen, there are significant differences in the types of fuel used in this industry in the two countries. For example, in the U.S. natural gas accounts for 34.5% of final energy use, while only accounting for 0.45% in China.

In addition to the share of fuels used directly in the iron and steel industry, the share of fuels used for power generation in each country is also an important factor, especially if the CO_2 emissions of the industry in two countries are compared. This becomes even more important because of the significant difference in the share of EAF steel production in China and the U.S. Since the share of EAF steel production in the U.S. is higher, the share of electricity use in total energy use is also higher compared to that of the Chinese iron and steel industry. In this case, the fuel share for the power generation in the country and as the result the emission factor of the grid (kg CO_2/kWh) plays an important role when comparing the CO_2 emissions of the iron and steel industry in the two countries. However, it should be noted that the comparison of the CO_2 emissions is beyond the scope of this report.

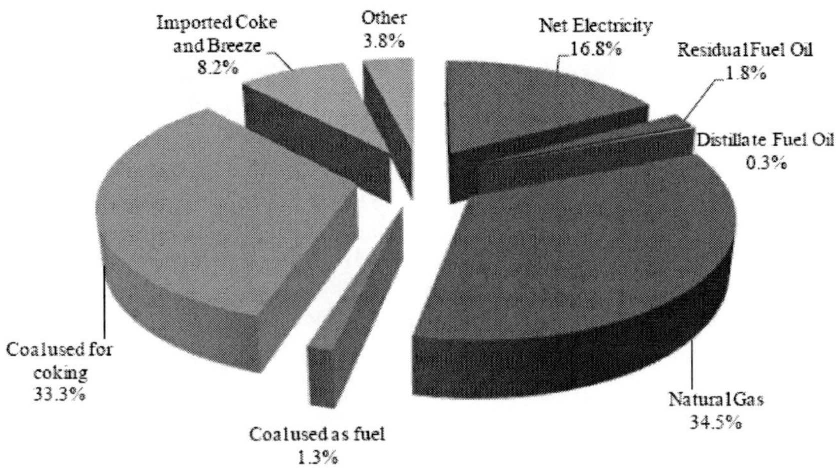

Source: U.S. DOE/EIA, 2010f Notes:

[1] These are fuel inputs to the U.S. iron and steel industry given the boundary defined in section 3.1.2. Therefore, the fuel conversion (e.g. from coal to coke) within the industry is not included in the figure above.

[2] Electricity is in final energy and is not converted to primary energy.

Figure 8. Total Energy Use (Fuel and Nonfuel) in the U.S. Iron and Steel Industry, 2006.

Figure 10 shows the fuel shares for electricity generation in the U.S. in 2006. Coal is the major source for power generation, accounting for about 49%, with around 20% of the electricity in the U.S. generated from natural gas, 19% from nuclear energy, and around 10% from hydroelectric and other sources of renewable energy. In contrast, 80% of electricity in China was generated by coal in 2006.

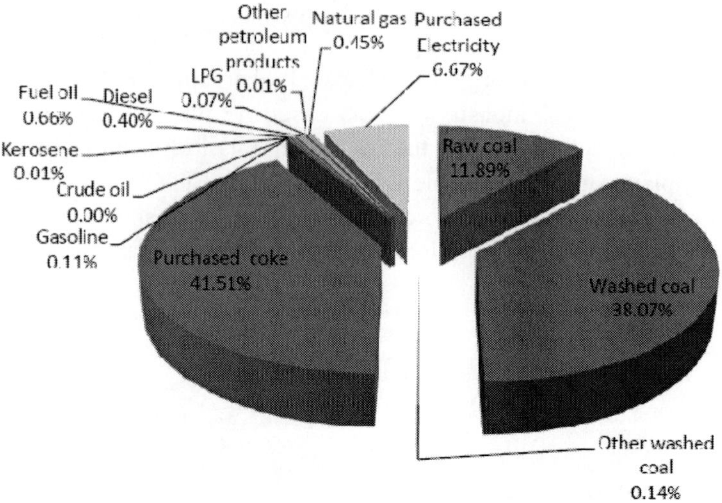

Source: China Iron and Steel Research Institute (CISRI) calculations based on NBS, 2010a.

Figure 9. China 2006 Primary Energy Use (Fuel and Non-fuel) of Iron and Steel Industry.

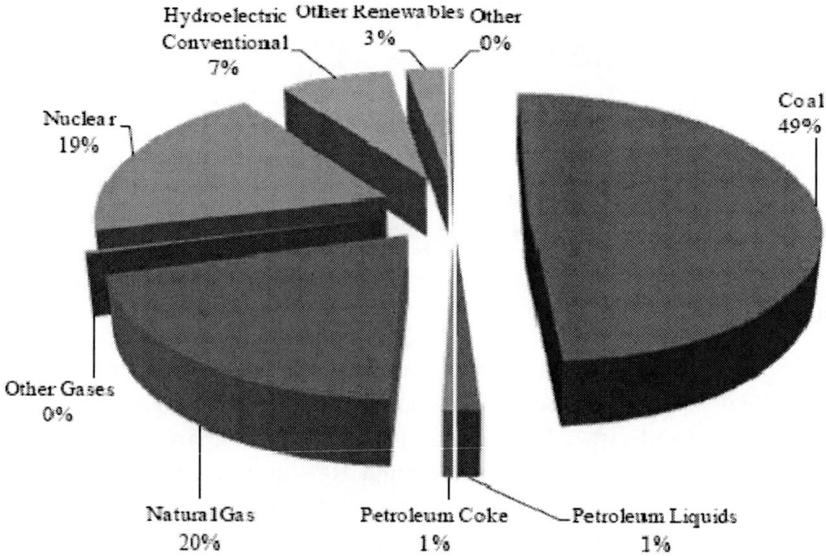

Source: U.S. DOE/EIA, 2010g.

Figure 10. Electricity Generation Fuel Shares in the U.S. in 2006.

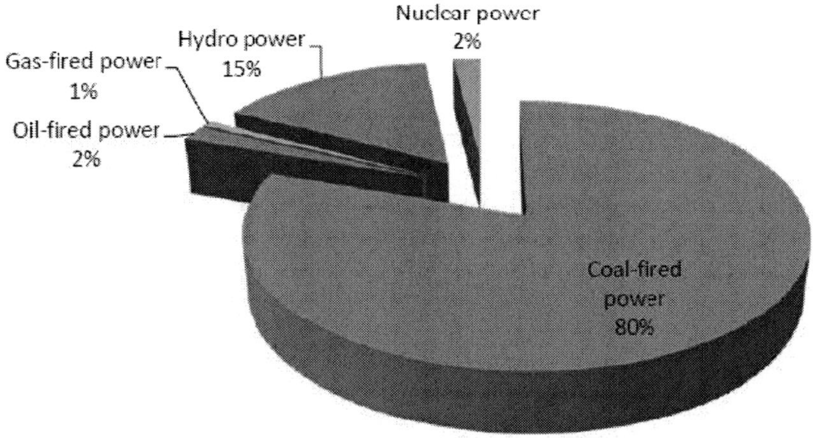

Source: NBS, 2010a.

Figure 11. Electricity Generation Fuel Shares in China in 2006.

The comparison of fuel shares in US and China shows that the share of fossil-fuel has a higher percentage in both countries, i.e., 83% in China (the sum of coal, oil and gas), and 71% in the U.S. (the sum of coal, natural gas, petroleum coking and oil). Power generation efficiency (coal consumption of net power supply, including plant self use) in both countries are similar.

5.4.3. Product Mix

Various final steel products have different energy requirements in the rolling/casting/finishing processes. Therefore, the final product mix is another key variable that should be considered when reviewing the results of the energy intensity comparison of the two countries. Table 40 and Table 41 show the final product mix of the iron and steel industry in the U.S. and China, respectively.

5.4.4. Penetration of Energy-Efficient Technologies and Practices

Data on penetration of energy-efficient technologies and practices in China and the U.S. are not fully comparable. The types of information available regarding the penetration of energy-efficiency technologies in these two countries are different. Therefore, a direct comparison of the penetration of certain technologies is not possible. The only technology that can be compared directly is the penetration of EAF steel production technology that was presented above.

Table 40. Final Product Mix of the Iron and Steel Industry in the U.S. in 2005 and 2006

	Quantity (thousand metric tons)		Percentage	
	2005	2006	2005	2006
Shipments by steel type:				
Carbon steel	88,800	92,700	93.25	93.33
Alloy steel	4,700	4,740	4.94	4.77
Stainless steel	1,730	1,890	1.81	1.90
Total	95,200	99,300	100.00	100.00
Steel mill products:				
Ingots, blooms, billets and slabs	1,160	1,500	1.22	1.51
Wire rods	1,760	1,840	1.84	1.86
Structural shapes, heavy	6,730	7,220	7.07	7.26
Steel piling	582	582	0.61	0.59
Plates, cut lengths	5,830	6,490	6.12	6.53
Plates, in coils	3,470	3,150	3.64	3.17
Rails	633	726	0.66	0.73
Railroad accessories	196	225	0.21	0.23
Bars, hot-rolled	6,300	7,060	6.61	7.11
Bars, light-shaped	1,470	1,490	1.54	1.50
Bars, reinforcing	6,350	6,830	6.67	6.87
Bars, cold finished	1,470	1,420	1.55	1.43
Tool steel	20	16	0.02	0.02
Pipe and tubing, standard pipe	1,030	1,140	1.08	1.15
Pipe and tubing, oil country goods	2,030	2,100	2.14	2.12
Pipe and tubing, line pipe	366	503	0.38	0.51
Pipe and tubing, mechanical tubing	998	966	1.05	0.97
Pipe and tubing, pressure tubing	38	43	0.04	0.04
Pipe and tubing, stainless	14	15	0.01	0.01
Pipe and tubing, structural	130	141	0.14	0.14
Pipe for piling	11	10	0.01	0.01
Wire	607	681	0.64	0.69
Tin mill products, blackplate	163	136	0.17	0.14
Tin mill products, tinplate	1,860	1,840	1.95	1.85
Tin mill products, tin-free steel	493	544	0.52	0.55
Tin mill products, tin coated sheets	96	93	0.10	0.09
Sheets, hot-rolled	20,300	19,400	21.30	19.55
Sheets, cold-rolled	11,600	12,000	12.19	12.13
Sheets and strip, hot dip galvanized	13,600	14,800	14.29	14.94
Sheets and strip, electrogalvanized	2,240	2,370	2.35	2.39
Sheets and strip, other metallic coated	1,710	1,890	1.80	1.90
Sheets and strip, electrical	424	481	0.45	0.48
Strip, hot rolled	36	38	0.04	0.04
Strip, cold rolled	1,520	1,500	1.59	1.51
Total	95,200	99,300	100.00	100.00

Source: U.S. Geological Survey (USGS), 2008.

The application of energy-efficient technologies depends on various factors such as raw materials, energy sources, cost, availability, product mix, and the regulatory regime in the country.

Penetration of energy-efficient technologies and practices in the U.S. iron and steel industry In 2006, out of 348 establishments[21] in the U.S. iron and steel industry, only 17 use cogeneration technology[22] (U.S. DOE/EIA, 2010k). Also, 40 establishments report using computer control of the building-wide environment (e.g. space-heating equipment, cooling equipment, lights), 199 use adjustable speed motors, and 88 use oxy-fuel firing (U.S. DOE/EIA, 2010j). In 2002, plants provided additional information regarding their energy management activities (Table 42).

Table 41. Final Product Mix of the Iron and Steel Industry in the China in 2005 and 2006

Year	2005	2006
Steel production, Mt/year	325.90	390.77
Bars, Mt/year	187.83	217.59
Plates, Mt/year	109.45	135.36
Pipes and tubes, Mt/year	26.14	35.52
Percentage of bars, %	57.63	55.68
Percentage of plates, %	33.59	34.64
Percentage of pipes and tubes, %	8.02	9.09

Source: Editorial Board of the China Iron and Steel Industry Association Yearbook, 2006.

Table 42. Energy Management in U.S. Iron and Steel Industry (NAICS 331111)

Item	# of plants (a, b)
Participation in One or More of the Following Types of Activities	412
Energy Audits	287
Direct Electricity Load Control	214
Special Rate Schedule (c)	118
Standby Generation Program	55
Equipment Rebates	30
Power Factor Correction or Improvement	213
Interval Metering (to manage energy use for programs such as real-time pricing)	69
U.S. Environmental Protection Agency's Energy Star Program	-
U.S. Environmental Protection Agency's Green Lights Program	12
Other Federally Sponsored Energy or Environmental Management Program	5
Equipment Installation or Retrofit for the Primary Purpose of Improving Energy Efficiency Affecting:	
Steam Production/System	172

Item	# of plants (a, b)
Compressed Air Systems	110
Direct/Indirect Process Heating	131
Direct Process Cooling, Refrigeration	47
Direct Machine Drive	199
Facility HVAC	150
Facility Lighting	148
Equipment Installation or Retrofit for the Primary Purpose of Using a Different Energy Source	49
Other	7
Full-Time Energy Manager	17

[a] 2006 information is not available.
[b] This count includes only those establishments that reported this activity in 2002.
[c] Examples are interruptible rates and time-of-use rates. Source: U.S. DOE/EIA, 2010i

Penetration of Energy-Efficient Technologies and Practices in the China's Iron and Steel Industry

With the rapid development of China's iron and steel industry, energy-efficient technologies and processes have also greatly improved. Penetration of equipment as well as technologies for waste heat and waste energy recycling has increased. The main technologies that are utilized include: coke dry quenching (CDQ) for the coking process, top-pressure recovery (TRT) for blast furnaces, pulverized coal injection, and comprehensive energy management and utilization technologies. Other technologies, such as low temperature waste heat recovery, are also being used gradually. The application and popularization of these energy-saving technologies have helped China to improve its energy efficiency in iron and steel industry. Many Chinese steel companies have benefited from the Kyoto Protocol's Clean Development Mechanism for additional funding for implementing CDQ and TRT projects in their plants, whereas the U.S. steel plants did not have such opportunities for outside investment-related funding.

1) CDQ and TRT Technologies

Figure 12 shows the penetration levels of CDQ and TRT in China's iron and steel industry since the 1990s, showing that they have increased significantly in recent years. Both CDQ and TRT have great energy-saving impacts. For example, CDQ can recycle more than 80% of the sensible heat of the heated coke. By quenching every ton of coke, it can recycle 0.45~0.6 ton of steam (at 4.5 MPa) on average (Shangguan, et al., 2009).The recycled steam can be fed into the streaming pipelines directly, or can be used for power generation as well. If using pure condensing steam turbines, on average 95~110kWh of electricity can be generated for quenching every ton of coke.

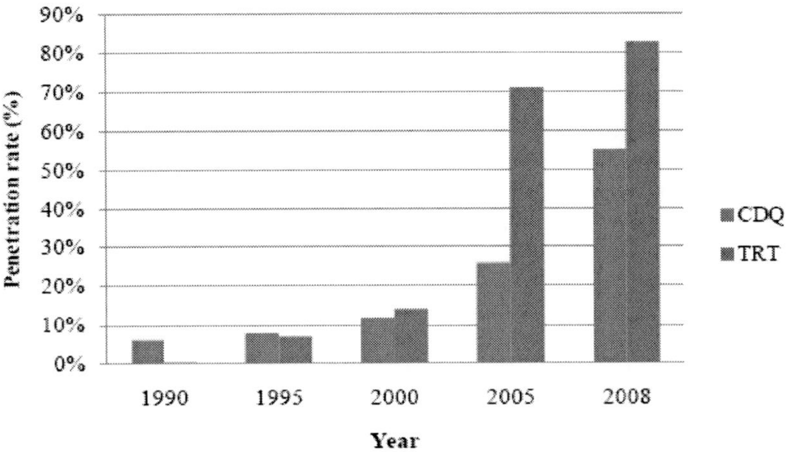

Sources: Gao and Gao, 2004; ISD-CISIA and ISIMI, 1991; Zheng, et al., 2005; Shan, 2008; Xue, 2007; EDRC, 2009; ISD-CISIA and ISIMI, 1996; Zhao and Jiang, 1999; EDRC, 2006; Zhang, 2006. Note: Penetration ratio of CDQ is the ratio at internal coking factories of steel mills.

Figure 12. Use of CDQ and TRT in China's Steel Industry.

TRT can recycle large amounts of fuel to produce electricity without consuming any fuel. According to statistical reports, if operated under good conditions, it can recycle 25~50 kWh per ton of steel, which can meet 30% electricity demand of the blast furnace. From 2000 to 2007, the number of blast furnaces with TRT in the Chinese steel plants increased from 33 to more than 400. By the end of 2007, all the blast furnaces with a capacity larger than 2000 m^3 are equipped with TRT, and 95% of the blast furnaces with a capacity larger than 1000 m^3 have TRT as well.

In addition, all of the TRT on blast furnaces (smaller than 1000 m^3) are dry-dust removal. Dry-dust removal has also been promoted into larger blast furnace (larger than 1000m^3). For instance, the TRT on two large blast furnaces (5500 m^3) in Tangshan Steel Mill (in Cao Pei Dian, China) utilize dry-dust removal. TRT with dry-dust removal can improve its efficiency by 30~40%, if compared to wet-dust removal, and reaches to 54kWh/t of steel (Shangguan, et al., 2009; ECERTF, 2008).

Considering the scale of China's iron and steel industry, the energy-saving impacts of both CDQ and TRT are significant. These technologies can contribute greatly to energy conservation and CO_2 emission reduction in the Chinese iron and steel industry.

2) Pulverized Coal Injection

Pulverized coal injection can reduce coke consumption in the blast furnace. Recently, the level of pulverized coal injection in the Chinese iron and steel industry has increased to 134 kg/t iron, which reaches the advanced international level (world average: 125 kg/t iron),[23] as presented in Figure 13.

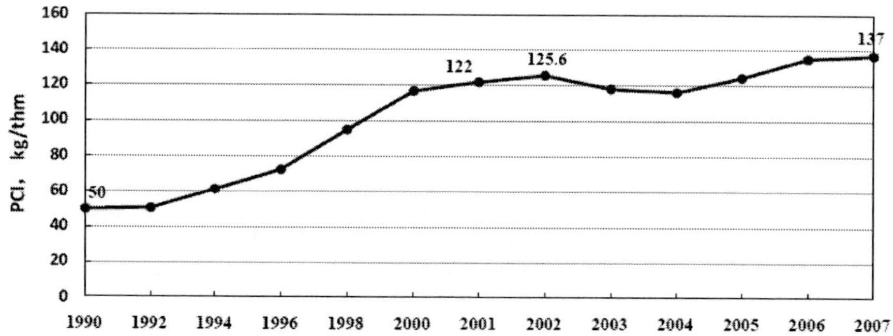

Source: Yin R, 2009.

Figure 13. PCI (Pulverized Coal Injection) Use in the Chinese Steel Industry, 1990-2007.

3) Continuous Casting

The ratios of continuous casting in the U.S. and China since 1990s are displayed in Figure 14 and Figure 15. Figure 14 shows that the continuous casting ratio in the U.S. had already reached a high level in the 1990s (about 76% in 1991). But the ratio in China during the same period was low, only 25% in 1990. With the development of China's iron and steel industry, the ratio of continuous casting increased rapidly, reaching 86.97% in 2000, and 98.57% in 2006. Currently, the continuous casting ratios in China and the U.S. are very close, with 97% in the U.S. in 2006. The improvement of the continuous casting ratio in China has had positive impacts on energy conservation and emission reduction in the Chinese iron and steel industry.

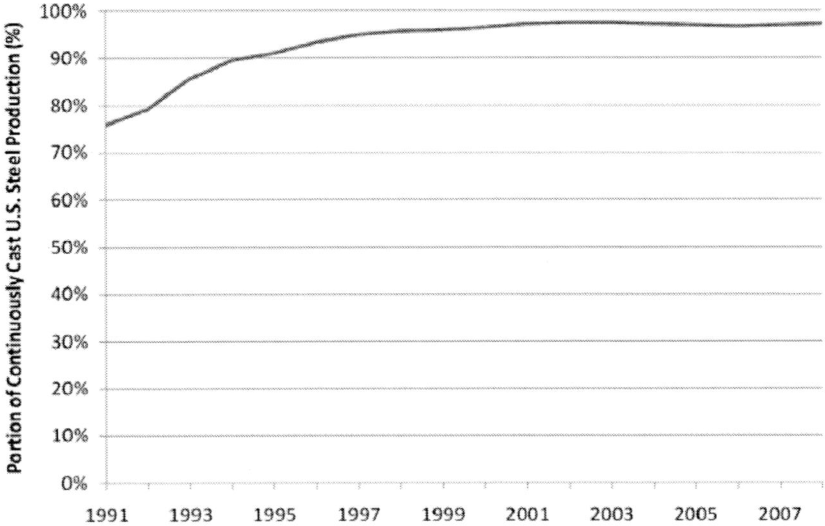

Source: USGS, 2010b.

Figure 14. Portion of Continuously-Cast U.S. Steel Production (1991-2008).

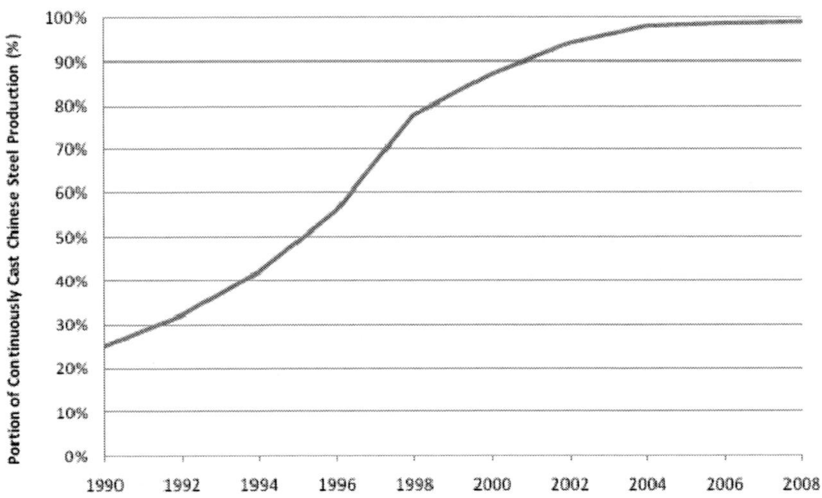

Source: Yin R, 2009.

Figure 15. Portion of Continuously-Cast Chinese Steel Production (1990-2008).

The analysis on the penetration and energy savings of energy-efficient technologies shows that the U.S. and China have their own characteristics in applying energy-saving technologies.

Energy management related technologies are emphasized more in the US. China has adopted more waste heat/energy recovery technologies, in addition to energy management related technologies.

5.4.5. Scale of Equipment

The U.S. steel industry is characterized by consolidated, large-scale integrated steel producers and largely fragmented, mini-mill secondary producers. Figure 16 illustrates the distribution of self-registered U.S. steel production facilities by annual capacity. The average capacity of integrated BOF plants in the registry was 2.9 Mt per year and 0.93 Mt for the EAF plants.

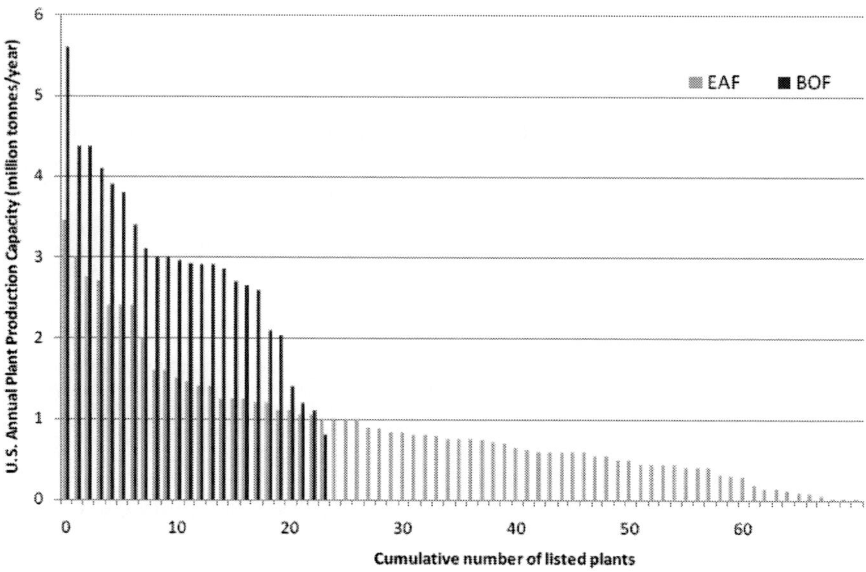

Source: Association for Iron and Steel Technology (AIST), 2008.

Figure 16. Distribution of Listed US Plant Facilities by Production Capacity (2007).

Overall, the Chinese iron and steel industry still has many small and inefficient enterprises or plants. There are many different types of steel enterprises in China as well, including large-scaled integrated steel enterprises, independent rolling enterprises, and even independent iron-making enterprises. Thus, the total number of iron and steel enterprises in China at the national level is quite large, and it is almost impossible to have the production and capacity information for every enterprise. But as the production from the key enterprises (medium-and-large sized) represents 82.9% of the national crude steel production (421 Mt in 2006), these plants can represent the level of major production equipment.

In 2006, China had 85 key medium-and-large sized enterprises, with a total crude steel production of 349 Mt. The average annual production capacity of each key enterprise was 4.1 Mt. From the perspective of average annual production capacity, China is larger than the U.S. Since 2006, China has been implementing a policy focused on phasing out inefficient capacities in energy intensive sectors. The overall level of the Chinese iron and steel industry is increasing gradually, along with the implementation of the phasing-out policy. From 2006 to 2009, China has phased out 82 Mt of iron-making production capacity and 60 Mt of steel-making production capacity (MIIT, 2010).

The key issue in China is the large share of small blast furnaces. The total number of blast furnaces in key enterprises in China in 2007 was 483 (EBCISIY, 1992-2009) among which 363 blast furnaces had a capacity under 1000 m^3. However, there were around 30 blast furnaces in the U.S. in 2007 with average capacity of 1500 m^3.

6. FINDINGS

A key finding of this analysis is that it is possible to develop a methodology in which the energy intensity of steel production of different countries can be compared. The methodology must clearly define the boundaries and energy conversion factors used in the analysis. The boundary definition must address how to account for imported and exported inputs and intermediate products.

Another key finding is that it is not possible to accurately compare the energy intensity of steel production of different countries without considering multiple scenarios. There is no single scenario that best compares different countries; each scenario presents different issues in terms of the accuracy and "fairness" of the comparison. For example, for this comparison of the U.S. and Chinese steel industries, the results change when the difference in production structure is taken into account when comparing the energy intensity values. For other countries, key differences might be found in the fuel or electricity conversion factors. Thus, it is necessary to present multiple scenarios to accurately convey the reasons behind the calculated energy intensities.

Thus, the base case 2006 comparison of U.S. and Chinese steel industries using identical boundary definitions for the steel industry, country-specific fuel and electricity conversion factors, and world average intensity values for net imported products, found that the final energy intensity of the Chinese steel industry is 23.11 GJ/t crude steel while that of the U.S. steel industry is 14.90 GJ/t crude steel (36% lower) and the primary energy intensity of the Chinese steel industry is 26.3 GJ/t crude steel while that of the U.S. steel industry is 19.98 GJ/t crude steel (24% lower).

A number of scenarios were also analyzed to identify which variables had the largest impact on the base case energy intensity values. First, the effect of the fuel conversion factors was examined in Scenario 1. Instead of using country-specific fuel conversion factors, IEA typical fuel conversion factors were used. All other variables were the same as in the base case. For this scenario, the final energy intensity of the Chinese steel industry is 22.65 GJ/t crude steel while that of the U.S. steel industry is 14.83 GJ/t crude steel (35% lower) and the primary energy intensity of the Chinese steel industry is 25.85 GJ/t crude steel while that of the U.S. steel industry is 19.91 GJ/t crude steel (23% lower). Thus, Scenario 1 shows that for these two countries, there is very little impact on the energy intensity values if standardized fuel conversion factors are used instead of country-specific factors.

Next, the impact of the electricity conversion factors was examined. For Scenario 2, world average electricity conversion factors were used instead of country-specific factors. All other variables are the same as in the base case. For this scenario, the final energy intensity of the Chinese steel industry is 23.11 GJ/t crude steel while that of the U.S. steel industry is 14.90 GJ/t crude steel (36% lower), which is identical to the base case. The primary energy intensity of the Chinese steel industry in this scenario is 25.77 GJ/t crude steel while that of the U.S. steel industry is 19.09 GJ/t crude steel (26% lower). Electricity conversion factors have more of an influence on the calculation of primary energy, which accounts for electricity generation, transmission, and distribution losses. Using the world average electricity values lowered the overall primary energy intensity in both countries, but the reduction was greatest on the U.S. value due to the larger share of EAF steel in the U.S.

A 3rd Scenario was developed to evaluate the combined impact of using average fuel and electricity conversion factors. For this scenario, the final energy intensity of the Chinese steel industry is 22.65 GJ/t crude steel while that of the U.S. steel industry is 14.83 GJ/t crude steel (35% lower) and the primary energy intensity of the Chinese steel industry is 25.32 GJ/t crude steel while that of the U.S. steel industry is 19.02 GJ/t crude steel (25% lower).

Comparison of Scenario 3 and the Base Scenario indicates that variation among country-specific fuel and electricity conversion factors does not account for a large portion of the energy intensity difference.

In order to evaluate the impact of the different structures of the steel industry in the two countries, Scenarios 4 and 5 were developed. Scenario 4 is the same as the base case, but each country's energy intensity is calculated using the EAF share in the other country. In 2006, the share of EAF steel production in total steel production was 10.5% in China and 56.9% in the U.S. When the U.S. energy intensity is calculated using the share of EAF steel production in China, the final energy intensity is 22.96 GJ/t crude steel and the primary energy intensity is 26.08 GJ/t crude steel. When the Chinese energy intensity is calculated using the share of EAF steel production in the U.S., the final energy intensity is 18.44 GJ/t crude steel and the primary energy intensity is 22.8 GJ/t crude steel. This calculation shows that the structure of the steel industry heavily influences the energy intensity and that if the U.S. steel industry was structured the same as the Chinese steel industry, the U.S. steel industry would be only 0.6% less energy-intensive than China in terms of the base case final energy and 0.84% less energy-intensive in terms of primary energy. It also shows that if the Chinese steel industry was structured the same as the U.S. steel industry, it would still be 19.2% more energy-intensive than the U.S. in terms of final energy and 12.37% more energy-intensive in terms of primary energy.

Scenario 5 is similar to Scenario 4, except that instead of country-specific fuel and electricity conversion factors, the IEA and worldsteel conversion factors used in Scenario 3 are used. Like Scenario 4, this scenario evaluates the impact of different structures on the energy intensity value.

When the U.S. energy intensity is calculated using the share of EAF steel production in China, the final energy intensity is 23.04 GJ/t crude steel and the primary energy intensity is 26.09 GJ/t crude steel. When the Chinese energy intensity is calculated using the share of EAF steel production in the U.S., the final energy intensity is 17.99 GJ/t crude steel and the primary energy intensity is 22.48 GJ/t crude steel. This calculation shows that the structure of the steel industry heavily influences the energy intensity and that if the U.S. steel industry was structured the same as the Chinese steel industry, the U.S. steel industry would be only 0.3% less energy-intensive than China in terms of final energy and the Chinese steel industry is 0.4% less energy-intensive than the U.S. in terms of primary energy. It also shows that if the Chinese steel industry was structured the same as the U.S. steel industry, it would still be 17.2% more energy intensive than the U.S. in terms of final energy and 7% more energy-intensive in terms of primary energy. These higher differences could be because of the fact that the energy intensity of EAF steel production in China is significantly higher than that of in the U.S. (see Table 39).

The final scenario developed for this analysis, Scenario 6, evaluated the impact of the different electricity final to primary conversion factors in each country. All other variables are the same as in the base case. This analysis reflects the difference in the generation, transmission, and distribution losses experienced in each country. As explained earlier, the final to primary conversion factors used in this analysis for 2006 are 0.376 kgce/kWh and 0.379 kgce/kWh for China and the U.S., respectively. This calculation shows that the there is little impact on the energy intensity value for each country if the final to primary conversion factor of the other country is used because the electricity conversion factors in two countries are similar.

Figure 17 shows the effect of various factors on the differences in energy intensity of iron and steel industry in China and the U.S. It should be noted that this figure is just for illustration purpose. The magnitude of each factor could not be determined because of the lack of data. However, it is found that the structural difference (i.e. difference in EAF ratio) between the steel industries in two countries has the highest contribution to the energy intensity difference.

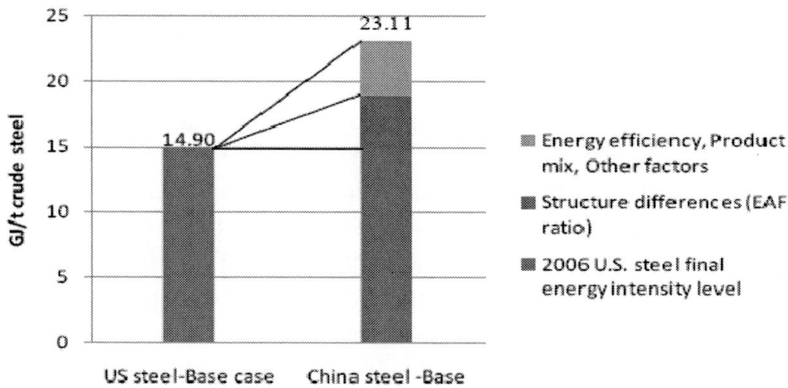

Note: This figure is the representative of the rough calculation for the contribution of different factors in the differences in energy intensity based on the scenario analysis done in this study. Further research is required to determine the exact magnitude of each factor.

Figure 17. Qualitative Effect of Key Factors on the Energy Intensity of Iron and Steel Production in China and the U.S.

7. Discussion, Policy Implications, and Future Research

7.1. Discussion

The findings presented above indicate that using a series of scenarios to evaluate the impact of different variables on the calculated energy intensity value for the steel industry in China and the U.S. can help to explain the observed differences in energy intensity. The results indicate that overall, the Chinese steel industry is more energy intensive than the U.S. steel industry. This is primarily, but not only, due to the difference in the structural composition of the two industries, with the U.S. producing a significantly greater share of the less energy-intensive EAF steel.

It is important to remember that China is at different development stage than the U.S., which contributes to differences in the efficiency of energy consumption. In the early stages of industrialization, economic growth in China has primarily depended on the development of energy-intensive industries.

The comparisons of energy consumption of different country's iron and steel industries need to take into account the impact of parameters such as fuel conversion coefficients, electricity conversion factors, and the EAF steel ratio. In addition, there are significant differences within China's iron and steel industry, with a number of inefficient "laggards" co-existing with top performers which represent the advanced level in the world (i.e. Baosteel Co., Ltd). If this analysis focused only on China's key steel enterprises, which account for more than 80% of crude steel production in China, the energy intensity values would most likely be lower than those for the entire country's steel industry (the energy intensity per ton steel of China's overall iron and steel industry is around 2.92 GJ/tonne crude steel more than that of the key steel enterprises).

On the other hand, much of China's steel production facilities have been constructed in recent years while most of the US steel production capacity is much older. As China phases out both smaller and older steel production capacity, the energy intensity of steel production in these two countries will converge slightly, but the more significant convergence happens when the share of EAF steel production in both countries converges as well. This may take a while until more scrap is available in China.

It is also important to note that during this analysis problems were discovered related to the availability and reliability of the data provided for the Chinese and U.S. iron and steel industries. Hence, both countries should make efforts to improve data collection and management.

7.2. Policy Implications

A number of policy implications result from this study. First, it should be noted that even with the use of a common methodology, it is difficult to provide policy-makers with a single energy intensity value for steel production for each country to be used to compare energy intensity across countries. Such values are often sought by policy-makers when making decisions related to energy, greenhouse gases, and competitiveness issues. This analysis illustrates that such a single indicator does not provide enough information to fully explain country-specific conditions.

In the case of the U.S. and China comparison, the key explanatory variable is the share of EAF steel. When comparing other countries, there may be other explanatory variables that are important. Thus, when providing policy-makers with a single indicator value for international comparisons, it is essential that explanatory variables also be evaluated and key results conveyed to policy-makers to accompany the single energy-intensity value.

This analysis also found that both countries could strengthen data collection for the iron and steel industry in order to both better understand trends in the industry and to more easily allow for the use of the methodology outlined in this report. For China, a number of adjustments had to be made to the data reported by the National Bureau of Statistics in order to make the comparison with the U.S. based on the outlined methodology. China could strengthen energy data collection and data management and consider adopting a system more similar to the system in the U.S. in which a census of manufacturing industries is conducted every four years. The U.S. could strengthen data collection and reporting related to facility and technology-level adoption of energy-efficient technologies and measures, such as CDQ and TRT.

During this study, it was also found that there is a lack of research-level communication and exchange between the Chinese and U.S. iron and steel industries, especially in the areas of iron and steel technologies, energy efficiency, and environmental protection. Since the Steel Sector Task Force of the Asia Pacific Partnership will move to the Global Superior Energy

Performance (GSEP) effort at the U.S. Department of Energy, there is potential for GSEP to establish a stronger, research-level sectoral communication platform among different countries, to learn from each other and to promote the mutual exchange of technical expertise as well as energy-efficient and environmental-protection technologies.

7.3. Future Research

Based on this analysis, there are three recommendations for future research. The first involves improving the analysis of the difference of application of energy-efficiency technology in iron and steel industry in both China and United States. This work may involve working with iron and steel research institutions or associations in both countries, or conducting surveys of key steel enterprises. Better information on both the penetration of energy-efficiency technologies as well as the barriers to the implementation of such options would be useful for both countries.

The second future area of research involves extending the current methodology and analysis to cover energy-related carbon dioxide emissions from steel production. Both the U.S. and China have committed to 2020 greenhouse gas emissions reduction targets under the Copenhagen Accord. In addition, China's 12th Five Year Plan includes a carbon dioxide emission reduction target of reduction of CO_2 emissions per unit of GDP by 17% between 2011 and 2015. Such an analysis will require additional methodological definitions and data, but could provide interesting and useful comparative information.

Finally, the third area of future research involved extending this analysis to include other key steel producing countries, such as Japan, India, Germany, and Brazil. Policymakers in both China and the U.S. may gain additional insights in how to improve domestic energy efficiency through further comparison with other countries which identifies not only the differences in energy intensity, but the key variables that explain the differences.

Different regulatory regimes in China and the U.S. can also influence the energy intensity of the steel industry in these two countries. For example, compliance with environmental regulations sometimes causes significant energy expenditures (e.g. in baghouse fans and filtration systems). Environmental regulations in the U.S. and some other developed countries are stringently enforced, while this may not always be the case in China. While the detailed comparison of the environmental regulations for the steel industry in China and the U.S. and quantifying the amount of energy used to meet those environmental regulations are beyond the scope of this study, this could be a subject for future research.

ACKNOWLEDGMENTS

This work was supported by the China Sustainable Energy Program of the Energy Foundation through the U.S. Department of Energy under Contract No. DE-AC02-05CH11231. The authors gratefully acknowledge the insightful review comments and suggestions of Ernst Worrell and Deger Saygin of Utrecht University, Christopher Weber of Science and Technology Policy Institute, Joan Pelligrino and Keith Jamison of Energetics, Lawrence Kavanagh of American Iron and Steel Institute, Henk Reimink of worldsteel Association, and Wang Yanjia of Tsinghua University. At LBNL, the authors gratefully acknowledge Hongyou Lu and Cecilia Fino-Chen for their valuable research assistance on this project.

REFERENCES

American Iron and Steel Institute (AISI), 2005. *Saving One Barrel of Oil per Tonne (SOBOT).* http://www.steel.org/AM/Template.cfm?Section = Articles7&TEMPLATE=/CM/ContentDisplay.cfm&CONTENTID= 12358.

American Iron and Steel Institute (AISI). 2008. *2007 Annual Statistical Report.* Washington DC.

Andersen JP and Hyman B. 2001. "Energy and material flow models for the US steel industry," *Energy* 26, no. 2 (February 2001): 137-159.

Anhua, Z. and Xingshu, Z. 2006. *Efficiency Improvement and Energy Conservation in China's Power Industry.* Available at: http://www.dfld.de/Presse/PMitt/2006/061030cI.pdf

Association for Iron and Steel Technology (AIST). 2010a. 2010 *North American BOF Roundup.* Iron & Steel Technology. November.

Association for Iron and Steel Technology (AIST). 2010b. 2010 *EAF Roundup.* Iron & Steel Technology. February.

Association for Iron and Steel Technology (AIST). 2008. *Directory of Iron and Steel Plants*. Warrendale, PA: AIST.

China Iron & Steel Association (CISA). 2003. *Comprehensive Statistical Data on 50 years of China's iron & Steel Industry*. Beijing: China Metallurgical Industry Press.

Chukwuleke OP, et al. 2009. "Shift From Coke to Coal Using Direct Reduction Method and Challenges". *Journal of Iron and Steel Research International*. 2009. 16(2): 01-05

China Iron and Steel Research Institute and China Iron and Steel Association (CISRI and CISA), 2010, "Energy Limit Values for the Steel Industry: National Standard and Product Application Guide (in Chinese)." Beijing: China Standards Press, June 2010.

Economic Development Research Center of Metallurgical Industry (EDRC). 2006. China Iron and Steel Industry Yearbook 2006. Beijing, China. [in Chinese]

Economic Development Research Center of Metallurgical Industry (EDRC), 2009. *China Iron and Steel Industry Yearbook 2009*. [in Chinese]

Editorial Board of China Iron and Steel Industry Yearbook (EBCISIY). 1992-2009. *China Iron and Steel Industry Yearbook*. Beijing, China [in Chinese]

Energetics, Inc. 2004. *Steel Industry Energy Bandwidth Study*. Available at: http://www1.eere.energy.gov/industry/steel/pdfs/steelbandwidthstudy.pdf

Energy Conservation and Emissions Reduction Task Force of China Iron and Steel Industry Association (ECERTF). 2008. "Directions and Measures of Energy Conservation and Emission Reduction in China's Iron and Steel Industry," *China Iron and Steel Industry*, 2008, (10): 39-44 [in Chinese]

Gao JY and Gao XX. 2004. Overview and Outlook of Global CDQ Technology Development. *Metallurgical Information Review,* 2004(3): 5-7 [in Chinese]

Information Statistics Department of China Iron and Steel Industry Association (ISD-CISIA), & Information and Standard Institute of Metallurgical Industry (ISIMI). 1991. *China Iron and Steel Statistics 1991*. [in Chinese]

Information Statistics Department of China Iron and Steel Industry Association (ISD-CISIA), & Information and Standard Institute of Metallurgical Industry (ISIMI). 1996. *China Iron and Steel Statistics1996*. [in Chinese]

International Energy Agency (IEA). 2005. *Energy Statistics Manual*. Paris: IEA. http://www.iea.org/textbase/nppdf/free/2005/statistics_manual.pdf

International Energy Agency (IEA). 2007. *Tracking Industrial Energy Efficiency and CO2 Emissions.* Paris: IEA. http://www.iea.org/w /bookshop/add.aspx?id=298

International Energy Agency (IEA). 2008a. *Energy Balances of Non-OECD Countries - 2008 Edition.* Paris: IEA.

International Energy Agency (IEA). 2008b. *Energy Balances of OECD Countries - 2008 Edition.* Paris: IEA.

International Energy Agency (IEA). 2008c. *Key World Energy Statistics 2008.* Paris: IEA. http://www.iea.org/textbase/nppdf/free/2008/key_stats_ 2008. pdf

International Energy Agency (IEA). 2010. *Energy Technology Perspectives 2010: Scenarios and Strategies to 2050.* Paris: IEA.

Kim Y and Worrell E. 2002. "International comparison of CO2 emission trends in the iron and steel industry," *Energy Policy* 30, no. 10 (August 2002): 827-838.

Kirschen M. et al. 2009. "Energy efficiency and the influence of gas burners to the energy related carbon dioxide emissions of electric arc furnaces in steel industry," *Energy* 34 (2009) 1065–1072.

Metals Processing Advisor. 2008. *Iron and Steel Overview: Energy Consumption.*
http://www.energysolutionscenter.org/heattreat/metalsadvisor/iron_and_st eel/overview/ energy_consumption.htm

Ministry of Industry and Information Technology (MIIT). 2010. Phasing Out Backward Capacities for 11th FYP. http://www.miit.gov.cn/n11293472 /n11293832/n13095885/13232432.html

National Bureau of Statistics (NBS). 2007. *China Energy Statistics Yearbook 2006*. Beijing: China Statistics Press.

National Bureau of Statistics (NBS). 2008. *China Energy Statistics Yearbook 2007*. Beijing: China Statistics Press.

National Bureau of Statistics (NBS). 2010a. *China Energy Statistical Yearbook 2009*. Beijing: China Statistics Press.

National Bureau of Statistics (NBS). 2010b. *China Industry and Product Categories.*
http://www.stats.gov.cn/tjbz/hyflbz/P020060711389582037795.pdf

Research Institute of Innovative Technology for the Earth (RITE), 2009. *Estimation of Energy Efficiency in 2005 (Steel Sector)*, RITE Systems Research Group.

Ruth M, et al. 2000. "Impacts of Market-based Climate Change Policy on the US Iron and Steel Industry," *Energy Sources*, Vol. 22, No. 3, pp. 269 - 280.

Shan SH. 2008. Promoting Energy Conservation and Emission Reduction with Green Steel. *Metallurgical Management*, 2008(6): 7-9 [in Chinese]

Shangguan FQ, Li XP, and Zhang CX. 2009. "Main Energy-Saving Measures in Iron and Steel Production and Analysis on Their CO2 Emission Reduction Potentials," *Metallurgical Energy*, 2009, 28(1): 3-7 [in Chinese]

Stubbles J. 2000. *Energy Use in the U.S. Steel Industry: A Historical Perspective and Future Opportunities*. U.S. Department of Energy, Office of Industrial Technologies, Washington, DC. http://www1.eere.energy.gov/industry/steel/pdfs/steel_energy_use.pdf

Tornell A. 1997. *Rational Atrophy: The US Steel Industry*. Working Paper 6084. Cambridge: National Bureau of Economic Research.

US Census Bureau. 2007. The North American Industry Classification System (NAICS). http://www.census.gov/cgi-bin/sssd/naics/naicsrch

U.S. Department of Energy, Energy Information Administration (U.S. DOE/EIA). 2007. *Annual Energy Review 2006, Appendix A: Thermal Unit Conversion Factors.* Washington, DC: EIA. http://www.eia.doe.gov/aer/

U.S. Department of Energy, Energy Information Administration (U.S. DOE/EIA). 2008. *United States Electricity Profile (2008 Edition)- Table 10. Supply and Disposition of Electricity, 1990 Through 2008.* Washington, DC: EIA. http://www.eia.doe.gov/cneaf/electricity/st_profiles/us.html

U.S. Department of Energy, Energy Information Administration (U.S. DOE/EIA). 2009. *Annual Energy Review 2008, Appendix A: Thermal Unit Conversion Factors*. Washington, DC: EIA. http://www.eia.doe.gov/aer/pdf/aer.pdf

U.S. Department of Energy, Energy Information Administration (U.S. DOE/EIA). 2010a. *Manufacturing Energy Consumption Survey (MECS)-* 2006 data (Table 1.2). Washington, DC: EIA. http://www.eia.doe.gov/emeu/mecs/mecs2006/2006tables.html

U.S. Department of Energy, Energy Information Administration (U.S. DOE/EIA). 2010b. *Manufacturing Energy Consumption Survey (MECS)-2006* (Table 1.1). Washington, DC: EIA. http://www.eia.doe.gov/emeu/mecs/mecs2006/2006tables.html

U.S. Department of Energy, Energy Information Administration (U.S. DOE/EIA). 2010c. *Manufacturing Energy Consumption Survey (MECS)-*

2006 (Table 3.1). Washington, DC: EIA. http://www.eia.doe.gov /emeu/mecs/mecs2006/2006tables.html

U.S. Department of Energy, Energy Information Administration (EIA). 2010d. *Manufacturing Energy Consumption Survey (MECS)-2006* (Table 2.1). Washington, DC: EIA. http://www.eia.doe.gov/emeu/mecs /mecs2006 /2006tables.html

U.S. Department of Energy, Energy Information Administration (U.S. DOE/EIA), 2010e. *Manufacturing Energy Consumption Survey (MECS)-2006* (Table 2.2). Washington, DC: EIA. http://www.eia.doe.gov/emeu /mecs/mecs2006/2006tables.html

U.S. Department of Energy, Energy Information Administration (U.S. DOE/EIA). 2010f. *Manufacturing Energy Consumption Survey (MECS)-2006* (Table 3.2). Washington, DC: EIA. http://www.eia.doe.gov/emeu /mecs/mecs2006/2006tables.html

U.S. Department of Energy, Energy Information Administration (U.S. DOE/EIA). 2010g. *Electric Power Industry 2008: Year in Review*. (Table 2.1 and Table 5.3). Washington, DC: EIA. http://www.eia.doe.gov/cneaf /electricity/epa/epa_sum.html

U.S. Department of Energy, Energy Information Administration (U.S. DOE/EIA). 2010h. Net Generation by Energy Source: Total (All Sectors). Washington, DC: EIA. http://www.eia.doe.gov/cneaf /electricity/epm /table1_1.html

U.S. Department of Energy, Energy Information Administration (U.S. DOE/EIA). 2010i. Manufacturing Energy Consumption Survey (MECS)-2002 data (Table 8.1). Washington, DC: EIA. http://www.eia.doe.gov /emeu/mecs/mecs2006/2006tables.html

U.S. Department of Energy, Energy Information Administration (U.S. DOE/EIA). 2010j. Manufacturing Energy Consumption Survey (MECS)-2006 data (Table 8.2). Washington, DC: EIA. http://www.eia.doe.gov /emeu/mecs/mecs2006/2006tables.html

U.S. Department of Energy, Energy Information Administration (U.S. DOE/EIA). 2010k. Manufacturing Energy Consumption Survey (MECS)-2006 data (Table 8.3). Washington, DC: EIA. http://www.eia.doe.gov /emeu/mecs/mecs2006/2006tables.html

U.S. Geological Survey *(*USGS*)*, 2004. *2003 Minerals Yearbook- Iron and Steel.* http://minerals.usgs.gov/minerals/pubs/commodity/iron_&_steel /festemyb03.pdf

U.S. Geological Survey (USGS), 2007. *2006 Minerals Yearbook- Lime.* http://minerals.usgs.gov/minerals/pubs/commodity/lime/myb1-2006-lime.pdf

U.S. Geological Survey (USGS), 2008. *2006 Minerals Yearbook- Iron and Steel.* http://minerals.er.usgs.gov/minerals/pubs/commodity/iron_&_steel /myb1-2006- feste.pdf

U.S. Geological Survey (USGS), 2010a. *Iron and Steel Statistics.* http://minerals.usgs.gov/ds/2005/140/ironsteel.pdf

U.S. Geological Survey (USGS). 2010b. *Iron and Steel Mineral Commodity Summary.* Various years. http://minerals.usgs.gov/minerals/pubs /commodity/iron_&_steel/

U.S. Geological Survey (USGS). 2011. *Iron and Steel Mineral Commodity Summary.* http://minerals.usgs.gov/minerals/pubs/commodity/iron_&_steel/mcs-2011-feste.pdf

Wang ZH. 1995. Moisture Adjustment of Feed Coal and Related Benefits. *Journal of Coal Chemical Industry*, 2005(3):46~54 [in Chinese]

World Steel Association (worldsteel). n.d. *CO2 Emissions Data Collection – User Guide, Version 6*, http://www.worldsteel.org/climatechange/files/2/2 /Data%20collection%20user%20guid e.pdf

World Steel Association (worldsteel). 2008a. Sustainability Report of the World Steel Industry.2008. http://www.worldsteel.org/pictures/publication files/Sustainability%20Report%202008_ English.pdf

World Steel Association (worldsteel). 2008b. World Steel Sustainability Indicator Methodology. http://www.worldsteel.org/pictures/storyfiles /SR08 %20mthodology.pdf

World Steel Association (worldsteel). 2008c. Fact Sheet - Energy. http://www.worldsteel.org/pictures/programfiles/Fact%20sheet_Energy. pdf

World Steel Association (worldsteel). 2009. Steel Statistical Yearbook 2008. http://www.worldsteel.org/pictures/publicationfiles/SSY2008[1].pdf

World Steel Association (worldsteel). 2010. *Steel in Figures.* http://www.worldsteel.org/?action=stats_search&keuze=steel&country=6 3&from=2009 &to=2009

World Steel Association (worldsteel), 2011. *Crude Statistics Total 2010.* http://www.worldsteel.org/?action=stats&type=steel&period=latest&mont h=13&year=2 010

Worrell E, Price L, Martin N, Farla J, Schaeffer R. 1997. "Energy intensity in the iron and steel industry: a comparison of physical and economic indicators," *Energy Policy* 25, no. 7 (June): 727-744.

Xue HF. 2007. The Urgent Need to Improve Recycling Energy Efficiency of Secondary Energy in Iron and Steel Industry. *China Environmental Newspaper*. 2007 [in Chinese]

Yin R. 2009. *Metallurgical Process Engineering (Version 2).* Beijing: Metallurgy Industry Press. [in Chinese]

Yin RY, Zhang CX, QI YH, et al. 2003. "Greening in the Iron & Steel Industry," *Iron & Steel* (z1): 135~138 [in Chinese]

Zhang CX. 2006. Energy Conservation and Emission Reduction in the Chinese Iron-Making System. *Steel*, 2006, 41(11) [in Chinese]

Zhang CX, Shangguan FQ, Changqing H. 2010. "Steel Process Structure and Its Impact on CO_2 Emission," *Iron & Steel*, 2010, 45(5):10-15 [in Chinese]

Zhao P and Jiang HH. 1999. *Analysis of Energy-Saving Technologies in Iron and Steel Industry.* Metallurgical Industry Press. Beijing, China. [in Chinese]

Zheng WH and Yu ZD. 2005. Current New Technologies of Coking Production in China. *Proceedings of 2005 China Iron and Steel Annual Conference*, 135~140. Beijing: Metallurgical Industry Press. [in Chinese]

APPENDIX

Table 43. Unit conversion factors

From this unit	Multiply by	To get this unit
1gce	0.0081	kwh
1MJ	0.0341208	kgce
1toe	1.4286	tce
1 million Btu	1055.056	MJ
1short ton	0.907	metric tonne
1 ft3	0.028316847	m3
1 barrel [US, petroleum]	158.987	liter
1lb	0.4536	kg

End Notes

[1]Note that the 2009 China Energy Statistical Yearbook lists total primary energy use for Smelting and Pressing of Ferrous Metals as 447 million tons of coal equivalent (Mtce) in 2006, thereby comprising 17% of total primary energy use for that year (NBS, 2010a). This also includes the energy use by facilities that belong to steel enterprises but are not part of the steel production process such as residential houses of the enterprises. The results of this report, which is focused solely on the energy used for iron and steel production and is thus less comprehensive than the Yearbook category of Smelting and Pressing of Ferrous Metals, is that this industry accounted for 13.6% of total primary energy use in China in 2006.

[2] Final (or site) electricity is the electricity consumed at the production facility. This value, however, does not include the primary energy used to generate, transmit, and distribute electricity to the site. To convert final electricity to primary energy, the average efficiency of power generation and transmission and distribution losses must be taken into account.

[3] To convert kgce to GJ, multiply by 0.02931 and to convert kgce to Million Btu, multiply by 0.02778.

[4] These enterprises are members of the China Iron and Steel Association. A list of these companies can be found at here: http://www.chinaisa.org.cn/index.php?id=298

[5] The key steel enterprises do not represent the total Chinese iron and steel industry; thus the energy intensity of the whole iron and steel industry in China would be different from what is presented above for the key steel enterprises. Throughout this report all the data presented are for the whole Chinese iron and steel industry unless it is mentioned otherwise.

[6] Note that USGS 2011 reports 2010 steel production of 90 Mt.

[7] Final (or site) electricity is the electricity consumed at the production facility. This value, however, does not include the primary energy used to generate, transmit, and distribute electricity to the site. To convert final electricity to primary energy, the average efficiency of power generation and transmission and distribution losses must be taken into account. This is discussed further below.

[8] Primary energy use is the sum of fuel use and the electricity use that is converted to primary energy by taking into account the power generation efficiency and transmission and distribution (T&D) losses. However, it should be noted that worldsteel does not take into account the T&D losses when converting electricity from final to primary energy.

[9] The values given in Table 8 are national grid T&D losses. It should be noted that sometimes plants obtain power at high or medium voltages directly from transmission lines; thus the distribution losses for them might be different.

[10] CISRI calculations from the "the norm of energy consumption per ton unit product of major procedures of crude steel manufacturing process" and "the norm of energy consumption per ton unit product of coke production" published by National Development and Reform Commission.

[11] Coal use for coking (nonfuel) accounted for about 95% of total coal use in the U.S iron and steel industry

[12] NBS (through personal communication) stated that the percentage was 21.61% in 2006.

[13] Personal communication with Zhang Chunxia from CISRI. December 2010.

[14] Total final energy intensity of the US iron and steel industry using the U.S. country-specific energy conversion factors for the purchased coke and auxiliary/intermediary products instead of worldsteel conversion factor would be 14.5 GJ/tonne crude steel, which is around 2.7% less than the intensity calculated using worldsteel conversion factors.

[15] This calculation will be included if the worldsteel conversion factor for electricity with T&D losses can be identified.

[16] Note, however, that even at this smaller overall share, China produced 44.22 Mt EAF crude steel compared to 56.10 Mt EAF crude steel produced in the U.S. in 2006.

[17] This calculation will be included if the worldsteel conversion factor for electricity with T&D losses can be identified.

[18] This calculation will be included if the worldsteel conversion factor for electricity with T&D losses can be identified.

[19] This calculation will be included if the worldsteel conversion factor for electricity with T&D losses can be identified.

[20] This calculation will be included if the worldsteel conversion factor for electricity with T&D losses can be identified.

[21] "Establishments" includes those units which reported any of the five energy-saving technologies listed by MECS in use any time in 2006, plus those units where usage of those technologies was not ascertained (U.S. DOE/EIA, 2010k).

[22] This count includes only those establishments that reported this cogeneration technology in use any time in 2006 (U.S. DOE/EIA, 2010k).

[23] The world average is 125 kg/t iron (IEA, 2007).

INDEX

D

E

S

T

U

V

W

Y